Neuroscience for Artificial Intelligence

Neuroscience for Artificial Intelligence

Huijue Jia

JENNY STANFORD
PUBLISHING

Published by

Jenny Stanford Publishing Pte. Ltd.
101 Thomson Road
#06-01, United Square
Singapore 307591

Email: editorial@jennystanford.com
Web: www.jennystanford.com

British Library Cataloguing-in-Publication Data
A catalogue record for this book is available from the British Library.

ISBN 978-981-4968-78-2 (Hardcover)
ISBN 978-1-003-41098-0 (eBook)

Contents

Preface

After the book on the human microbiome, and as I was relocating from BGI-Shenzhen to Fudan University and IPM-GBA, it occurred to me that I could and should write another book, crossing into neuroscience and computation. And as it turned out, the last chapter of this book became a step further from Chapter 6 of the microbiome book (ISBN: 978-0-323-91369-0) on causality. Not sure whether Prof. Judea Pearl will get to see this book.

Some of the most fascinating phenotypes of humans keep their mysteries in the brain. Many animals are very smart, but have their own evolutionary constraints and priorities.

With new technologies such as single-cell recordings, calcium imaging, and optogenetics, neuroscience is seeing a burst in new literature. AI researchers, however, would probably find these publications dauntingly difficult for them to read. Whereas prevalent hierarchical algorithms look like dividing annual goals into departments in a company (e.g., in Dr. Yan LeCun's book), many functions of the animal and human brain are more bottom up. I hope this book would be a good start for engineers and computer scientists to tackle problems from their angle. This is as yet a scientific discussion, and the social and philosophical implications had better not be over-interpreted.

From academician Xiongli Yang's lecture in Shanghai No. 3 Middle School, I mostly remembered the elegant 4-leaf fans hanging from the ceilings, a slide on potential gender differences in the brain, and that he only needs 6 hours of sleep every night. A few years later (almost 20 years ago), I tried to take a neuroscience class, but deselected the course after listening to the first class, thanks to the flexible system at Fudan University. It can be safely said that I have no formal training in neuroscience. I got to play with Mathematica models of neuronal firing in an elective class at Case Western Reserve University (and have to apologize for being unable to find the professor's name now).

And after finding excellent collaborations with doctors in China, especially Drs. Xiancang Ma, Feng Zhu, and Dr. Zhenxin Zhang, we (Ruijin Guo at BGI) have been doing some interesting work on neuropsychiatric and neurodegenerative diseases.

Thank you for being interested in this book. It has been great fun to piece together discrete information for a coherent and, to my knowledge, original picture. Before we completely understand how the brain works, it is probably a good idea to keep feeding it with new experiences.

Huijue Jia
Shanghai
February 2022

Acknowledgments

I sincerely thank the staff at Jenny Stanford Publishing, Jenny, Arvind, and probably many more people I didn't get to know, for giving me the opportunity to publish this book. I thank Arvind for inviting expert previews and Prof. Kasai and Prof. Schüz for their enthusiasm and rigor. I'm an expert in neither neuroscience nor AI. And I have to apologize for being unable to cite all the literature from the different fields of neuroscience. I thank Yanzheng Meng, Changxing Su, Fei Li, and Lilan Hao for improving the figures for publication and Yanmei Ju for help with some references. I thank my dear husband, Dong, for support and for inspiration and my parents, and teachers over the years, for having nurtured a young and always curious brain. Prof. Carlos Bustamante from Berkeley said during a luncheon before a seminar that as scientists we should always push the boundary, instead of staying safely behind.

January 2023

Chapter 1

Evolving under Constraints

Abstract

The human brain has about 10^{11} cells, consumes less energy than existing artificial intelligence setups, while doing much more. From an evolutionary point-of-view, the metabolic burden and the wiring complexity of the primate brain is rather large. New functions have to be accommodated with existing ones, for the better survival of a species. This Chapter serves as an introduction to the design principles of the brain, before we get into details in the subsequent chapters that can inspire new designs in artificial intelligence.

Keywords

Brain structure, brain evolution, limbic system, cortex, Brodmann's areas (BA), brain folds, brain metabolism, wiring cost, synapses, synapse density

1.1 An Evolutionary Continuum

In a human-centered view, everything else was seen as more primitive. For the brain, however, every species, as long as they can afford it, has probably continued to evolve the structures and functions for its own prosperity on this planet (Figs. 1.1 and 1.2).

Neuroscience for Artificial Intelligence
Huijue Jia
Copyright © 2023 Jenny Stanford Publishing Pte. Ltd.
ISBN 978-981-4968-78-2 (Hardcover), 978-1-003-41098-0 (eBook)
www.jennystanford.com

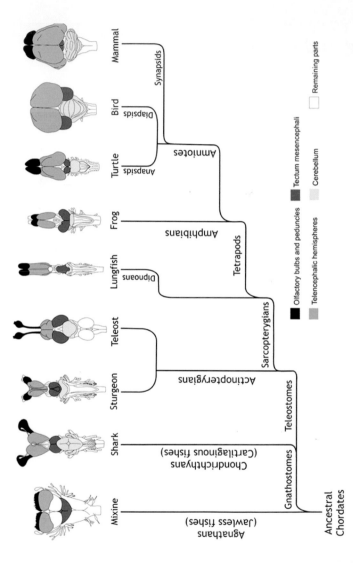

Figure 1.1 Evolution of the extant vertebrate brain. The drawings show a dorsal view of the brain of representative species of each phylogenetic group. In sharks and other fishes, the hindbrain is predominant, and the rest of the brain serves primarily to process sensory information. In amphibians and reptiles, the forebrain is far larger, and it contains a larger cerebrum devoted to associative activity. In birds, which evolved from reptiles, the cerebrum is even more pronounced. In mammals, the cerebrum covers the optic tectum and is the largest portion of the brain. The dominance of the cerebrum is greatest in humans, where it envelops much of the rest of the brain.

Credit: Fig. 2 of ref. 2.

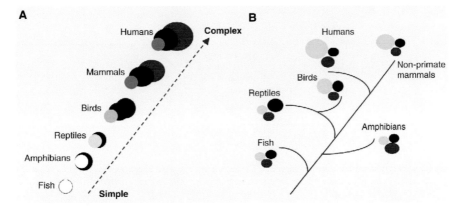

Figure 1.2 Schematic representation of two theories of brain evolution. (A) The outdated "scala naturae" theory, where evolution occurs in a linear, progressive fashion up a ladder in which "lower" (simple) species evolve into "higher" (complex) species; going from fish and amphibians at the bottom through reptiles and birds to primates and humans at the top. With respect to brain evolution, the increasing complexity resulting from climbing the ladder leads to the appearance of completely new areas which are then added onto old ones. Each color represents a different hypothetical brain region, either old or new. (B) The modern theory, where evolution is tree-like and new species evolve from older ancestral forms. With respect to brain evolution, complexity is derived from refining neural structures which are already present in ancestral forms, such that brain regions increase in size. There are no truly new brain areas, only elaborations of established regions. The colors represent different brain regions, but rather than new areas being added, evolutionarily old areas are increased or decreased in size (or complexity).

Credit: Fig. 1 of ref. 3.

The cerebral cortex has three layers in reptiles, while such lamination is lost in birds[1]—which we may find rather smart.

The human brain is not the largest. Mammalian brain size, and the volume of gray matter (a visible outer layer that is populated with neuronal cell bodies, Section 1.3), scale with body weight (Fig. 1.3). The volume of white matter (the brain mass enveloped by the gray matter layer, mostly consisting of axons and glial cells) increases faster than gray matter, as body weight increases[4] (Fig. 1.3).

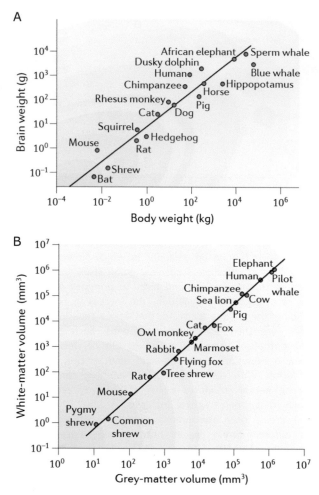

Figure 1.3 Allometric and fractal scaling of brains and human brain networks. (A) Larger organisms have larger brains[5]. (B) Larger brains have disproportionately more white matter than gray matter[6]. The line is the least squares fit, with a slope around 1.23 ± 0.01 (mean ± SD).

Credit: Fig. 1a,c of ref. 4. Part A was modified by ref. 4 with permission from ref. 4. Part B was modified by ref. 4 with permission from ref. 6.

1.2 Overall Structure of the Brain

We learned in elementary and secondary schools about the peripheral and the central nervous systems (Fig. 1.1). What we commonly refer to as the brain consists of the hindbrain (esp.

pons and cerebellum), midbrain and forebrain (esp. cerebrum, and interbrain including thalamus, hippocampus, etc.) (Figs. 1.1, 1.2, 1.4, and 1.5). The brainstem (the part between the interbrain and the spinal cord without the cerebellum, and

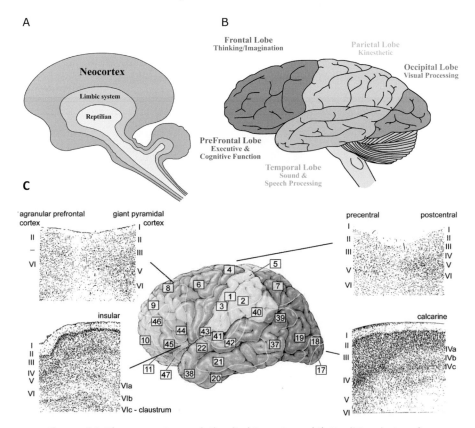

Figure 1.4 The neocortex and the limbic system. (A) Traditional view for the evolution of the neocortex on top of the limbic system. Birds, reptiles and fish do also have limbic structures such as the hippocampus[10, 11]. Cells in the neocortex do grow in an inside-out fashion along radial glial cells early in development after Cajal-Retzius cells establish the boundaries[12, 13]. (B) Cortical lobes and their major functions. (C) Classical cytoarchitectonic areas described by Brodmann[14] (area numbers (BA1-BA52) indicated in circles). Nissl staining show examples of transition between areas. For example, Layer VI of the insular cortex is split into sublayers VIa,b,c, the latter is then continuous with the adjacent claustrum (bottom left); a single layer IV in the secondary visual cortex splits into three sublayers IVa,b,c in the primary visual cortex (bottom right)[13].

Credit: Part A, B, Lilan Hao; Part C, Fig. 1A of ref. 13.

mostly included in the hindbrain) is responsible for lifeline functions such as breathing and heartbeats and is now known to play a key role in sleeping and dreaming (Chapter 6). The cerebral neocortex, also called the isocortex, was so named for its notable expansion in humans (esp. the prefrontal cortex, Figs. 1.4 and 1.5). The six layers of neurons in the human neocortex contrast the three layers in evolutionary older parts such as the hippocampus[1]. Retinoic acid (vitamin A) is key to the development of the human prefrontal cortex[7–9].

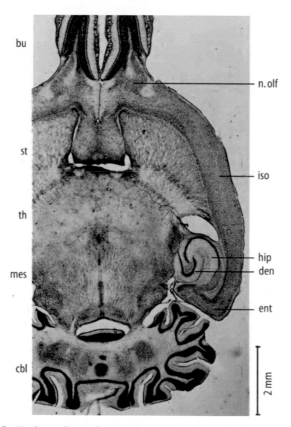

Figure 1.5 Horizontal Nissl-stained section through the mouse brain. Demarcation of the olfactory bulb from the cortex. Continuous transition of the cortex into the hippocampus (more in Chapter 4). bu, olfactory bulb; st, striatum; th, thalamus; mes, mesencephalon (midbrain); cbl, cerebellum; n. olf, olfactory nucleus; iso, isocortex (neocortex); hip, hippocampus; den, dentate gyrus, which is a separate sheet in hippocampus; ent, entorhinal cortex.

Credit: Fig. 2 from Chapter 2 of ref. 15.

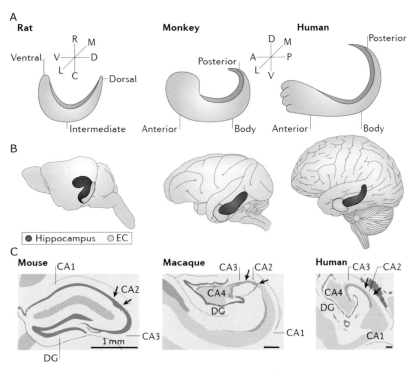

Figure 1.6 Cross-species comparison of hippocampal anatomy. (A) Schematic illustrations of the orientation of the hippocampal long axis in rats, macaque monkeys and humans. The longitudinal axis is described as ventrodorsal in rodents and as anteroposterior in primates (also referred to as rostrocaudal in nonhuman primates). There is currently no precise anatomical definition for a dorsal (or posterior) portion of the hippocampus relative to a ventral (or anterior) one, although in general, topologically, the former is positioned close to the retrosplenial cortex and the latter close to the amygdaloid complex. Note that a 90° rotation is required for the rat hippocampus to have the same orientation as that of primates. In primates, the anterior extreme is curved rostromedially to form the uncus. (B) The full long axis of the hippocampus (red) can be seen in brains of rats, macaque monkeys and humans, with the entorhinal cortex (EC) shown in blue. (C) Drawings of Nissl cross-sections of mouse, rhesus and human hippocampi. A, anterior; C, caudal; D, dorsal; DG, dentate gyrus; L, lateral; M, medial; P, posterior; R, rostral; V, ventral.

Credit: Fig. 1 of ref. 16. Panel A was adapted by ref. 16 with permission from ref. 17. Panel C was from ref. 18.

The limbic system is known as the part of the brain involved in behavioral and emotional responses, including behaviors such as feeding, fighting, reproduction and caring for the young

(Fig. 1.4). The amygdala, located in front of the hippocampus and behind the olfactory bulbs, is famous for emotional memories (arithmetics in Chapter 8). Despite being an evolutionarily ancient part of the brain[10,11], the hippocampus got into an orthogonal arrangement with many of the functional domains in the neocortex apparently only in primates (Fig. 1.6, cf. Figs. 1.4 and 1.7). I'll speculate more on such a topology as we discuss working memory in Chapter 8.

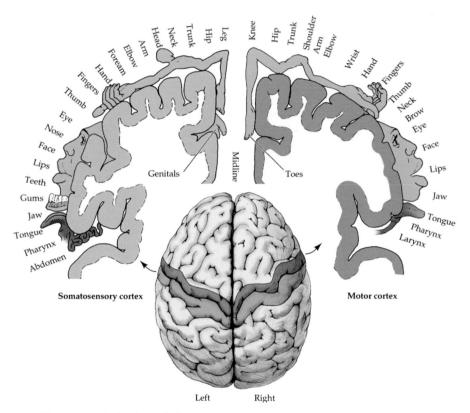

Figure 1.7 Projection of the human body in the sensory cortex and the motor cortex.

Credit: Fig. 2.13 of ref. 19.

The human somatosensory cortex and the motor cortex have large areas devoted to facial features, hands and feet (Figs. 1.4 and 1.7). The human visual cortex, located in the back of the brain (Fig. 1.4), typically use the central area to recognize faces (Section 2.4).

The 52 Brodmann areas of the human and monkey cerebral cortex, defined by Dr. Korbinian Brodmann over a century ago, is widely used when referring to the cortical regions[14] (Fig. 1.4C). Single neurons, while being units of computation, cannot yet be noninvasively recorded in healthy human volunteers. Functional magnetic resonance imaging (fMRI) which visualizes increased blood flow, is as yet widely used in studying activities in the brain of healthy volunteers.

Claustrum, a thin sheet located below the cortex (Fig. 1.4C, Fig. 7.15B), receives input from most regions of the cortex and makes long-range projections to most regions of the cortex[20-23], including the entorhinal cortex[23] (Fig. 1.6). Dr. Francis Crick has been fascinated by the claustrum until his last days, for its potential function in integrating conscious perception (Fig. 7.15).

Broca's area, together with nearby regions that constitute a "Broca's complex," is essential for rendition of a language[24, 25] (Fig. 1.4, Fig. 1.8, more on language in Chapter 8). The extended Wernicke's area functions in understanding a language (Fig. 1.8). Note its proximity to the auditory cortex, the entorhinal cortex and hippocampus (Figs. 1.4 and 1.6), which is likely a general organizing principle that saves wiring[4].

As brain space is so limited, boundaries between the functional domains might have been dynamic at least within a developmental time window[26]. The horizontal limb of the diagonal band of Broca has been found in rats to relay a linear relationship between locomotion speed and neuronal firing rates from the brainstem to the medial entorhinal cortex[27], so that rats can run and stop with proper coordination.

There is a general tradeoff between wiring cost (total volume of wiring) and path length, and the brain is already quite optimal[4]. As we'll discuss in the later chapters of this book, new dimensions are added on top of existing ones, which could then enjoy some independence in function.

Folding of linear polymers such as the chromatin in the cell nucleus has been shown to be a fractal globule that minimizes entanglement and allows dynamic remodeling of domains. So this higher energy state than an equilibrium globule is necessary for the flexible activities of chromatin[28].

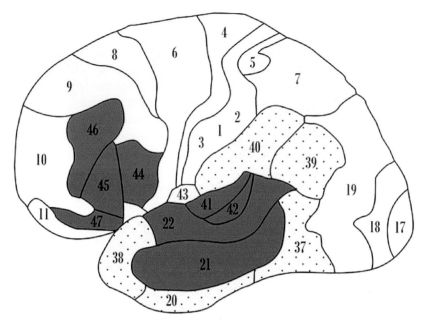

Figure 1.8 Brodmann's Areas (BA) involved in language. The frontal language area (Broca's complex: language production and grammar: BA44, BA45, BA46, BA47) also partially includes BA6 and extends subcortically to the basal ganglia. The posterior language area (language reception and understanding: lexical-semantic system) includes a core Wernicke's area (BA21, BA22, BA41, and BA42) and an "extended Wernicke's area" also including BA20, BA37, BA38, BA39, and BA40, involved in language associations. In addition to the well-recognized Broca's area (BA44 and BA45), there is a complex frontal-subcortical circuit involved in language production and grammar ("Broca's complex"). The insula (BA13, in between the Broca' complex and the Wernicke's area and not visible), probably plays a coordinating role in interconnecting these two brain language systems (lexical-semantic and grammatical).

Credit: Fig. 2 of ref. 24.

The brain cortex shows its characteristic folds (e.g., Figs. 1.4 and 1.7) due to faster growth of the cortex than the inside[29, 30], while also being confined in space. Folding of the gray matter in the cerebrum is two-dimensional, with the main axons about perpendicular to the cortex (Fig. 1.9), reenforced by other cells and interspersed by blood vessels (Chapter 6). During development, Cajal-Retzius cells establish the cortical boundaries, and radial glial cells form spokes that other cells can climb up with[12, 13].

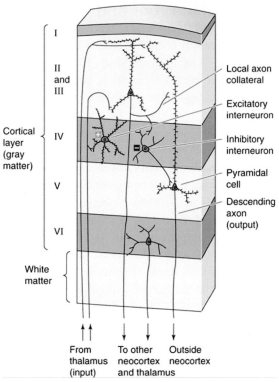

Figure 1.9 The six layers of the neocortex with major neurons. The cortex has six layers in mammals and three layers in reptiles (but birds lost such lamination)[1]. Layer I almost only contain dendrites from other layers, including Layer V[31]. A basic local circuit in the neocortex consists of inputs (e.g., afferent axons from the thalamus to Layer IV or Layer I), excitatory and inhibitory interneurons[32], and output neurons (e.g., pyramidal cells). Pyramidal cells, some interneurons (excitatory stellate cells, Martinotti cells), and Purkinje cells of the cerebellum (Chapter 3) have spines on their dendrites, which can constitute 75% of cortical synapses[15]. The dendritic spines can occupy a notable fraction of space, on the same order of magnitude as the dendritic shafts. While terms such as wiring and circuits have been popular in the literature, most of the adult cortex is an established framework of mostly excitatory synapses[15, 33]. But inhibitory synapses tend to be strategically placed at the main bifurcation of apical dendrites close to the border between L1 and L2 (ref. 33). All the spiny interneurons are excitatory glutamatergic neurons located in Layer IV that receive sensory input from the thalamus; the majority of the aspiny cortical interneurons are inhibitory GABAergic neurons located in all layers of the cortex[34]. This book will briefly mention development (Chapter 3) and then precede with the physical basis of memory (Chapters 4, 5).

Credit: Fig. 14.15 of ref. 35.

1.3 Number of Neurons and Their Connections

The mammalian cerebral cortex is a thin cortex of 1–3 mm thick (Fig. 1.9, Fig. 1.5). There are about 10^5 neurons per square millimeter (mm^2) of cortex[36, 37]. So a larger surface area (gray matter, Fig. 1.3) means more neuronal cells. For example, the human cerebral cortex has a surface area of 2600–4100 cm^2, 3400 times the area (and the number of neurons) of the mouse cerebral cortex[36].

More than a hundred years after the Spanish scientist Dr. Santiago Ramón y Cajal triumphed over the Reticular theory—the nervous system such as a brain being a continuous network—with clear staining of neuronal cells using the technique that he had meticulously improved from Dr. Camillo Golgi's staining method[38], the cellular and subcellular computational functions of the brain still appear underappreciated.

Researchers in the artificial intelligence (AI) field often quote the tens to hundreds of billions of neuronal cells in the brain (e.g., ref. 39). But as we shall see in this book, individual functions of the brain are often achieved with a small number of active cells, and heavily rely on bottom-up instead of top-down signals (Fig. 1.10).

A classic neuron outputs through its axon to the dendrites of another neuron (dendritic computation in Chapter 5) and the stable junction is called a synapse. According to meticulous analyses in mice, there are about 7.2×10^8 synapses per mm^3 of cortex, so the total number of cortical synapses is 10^{11} in mouse and ~10^{14} in human[15]. Neurons in a larger brain would have to connect with more neurons. On average each cortical synapse is 5 µm apart on axons and 0.5 µm apart on spiny dendrites[15], presenting plenty of opportunities for clustering of neighboring synapses on dendrites (Fig. 4.1, Chapter 5). Pyramidal neurons (in Layer II/III of the mice visual cortex) have a lower density of synapses within 50 µm of the cell body, whereas interneurons, especially basket cells (Section 6.3), have a higher density of synapses closer to the cell body[40].

A cortical synapse in mice is 320–380 nm (0.32–0.38 µm) wide[15] (Each synapse is an end junction between an axon and a dendritic spine or dendritic shaft, and the width of each synapse is therefore smaller than the diameter of a dendritic spine head, Chapter 5). For comparison, a standard *Escherichia coli* bacterial cell is a 2 μm rod of 0.5 μm diameter; a single eukaryotic ribosome is ~30 nm (we'll see them making proteins in dendrites, Fig. 5.5).

Although a minor group of cortical synapses[15,40,41] (Fig. 5.4), inhibitory synapses from inhibitory interneurons (Fig. 1.9) onto dendritic shafts appeared larger and more spaced out than excitatory synapses from pyramidal neurons onto dendritic spines[15,41]. And interneurons are indispensable for sensing speed (Chapter 7), memory (Chapter 5), associative learning (Chapter 9), and sleep (Chapter 6).

In the tightly packed cortex full of cell bodies, axons, dendritic spines and shafts, proximity appears to be a prerequisite for "wiring." Membrane surface available around an axon, especially within 5 μm, predicts a synaptic connection, according to three-dimensional reconstruction of electron microscopy images of layer 4 of the mice somatosensory cortex[42].

These initial setups are already quite different from the backpropagation algorithm[13,44] (Fig. 1.10) that is now commonly used in neural networks (Artificial Intelligence, AI), despite the apparent similarity in layering. Backpropagation can look like determining annual goals in a corporate group, divided among constituent companies and their departments. Each layer of the cortex is different (Fig. 1.9), also unlike the simple addition of hidden layers in AI.

Braitenberg and Schüz estimated that any two pyramidal neurons in the cortex (Fig. 1.9) are linked by two or three synapses[15,36], which is an impressive array of all possible associations (more on causal reasoning in Chapter 9). So the number of connections increases with the square of the number of cells (or subpopulations of cells), while the connections are relatively sparse, and activations are further gated by transmission probability (Chapter 5). Such sparsity saves energy and storage space, as well as ensuring robustness in the adult brain.

Opening of a single calcium channel due to change in membrane voltage could lead to vesicle release at some types of synapses, while other types of synapses may require many more calcium channels to open[45].

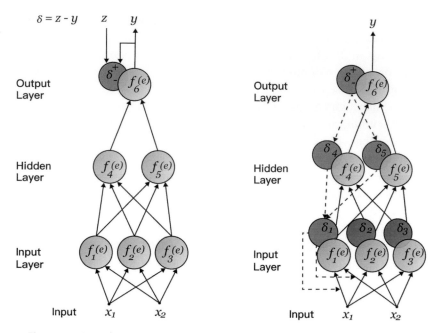

Figure 1.10 Backpropagation shown in a simple network of three layers. The weights propagate forward through the layers and are summed to predict *y*. Different between this *y* and the target *z*, shown as the error signal *δ*, is transmitted back through the layers, which cause re-weighting that runs forward for another comparison. The algorithm proceeds for a better fit by means of gradient descent.

Credit: Hao Zheng.

1.4 Fuel for the Brain

Glucose is fuel for the brain (Fig. 1.11). The human brain consumes 20% of the body's resting-state energy (half of which is used on the Na$^+$/K$^+$ pump) while being only about 2% of adult body weight[46, 47]. The brain metabolism of glucose is substantially decreased in Alzheimer's disease and its pre-stage, amnestic mild cognitive impairment (aMCI), with accumulated oxidative damage in a variety of molecules[48]. Age-related accumulation of the metabolite N^6-carboxymethyllysine in microglia (immune cells in the brain), contributed by microbes in the gut, increased oxidative stress and mitochondrial damage in the aging brain[49]. How such global trends are reflected in the metabolic burden of single neurons and nonneuronal cells remains to be investigated, which could potentially offer new insights for memory and sleep (Chapters 4, 5, 6). A hypertonic solution of sucrose or other pressure has been shown to induce release of vesicles at synapses[50].

After glucose is depleted during starvation/exercise, the liver produces ketone bodies as an alternative fuel especially for the brain. Ketone bodies such as β-hydroxybutyrate promotes brain-derived neurotrophic factor (BDNF) expression in the hippocampus via inactivation of histone deacetylases (HDACs)[51]. To rely on ketone bodies is like a battery-saving mode, but intermittent switching into it can help clean-up and restart the system. For example, a ketogenic diet protects against seizures and potentially other neurologic diseases through modulation of the microbiome[52, 53].

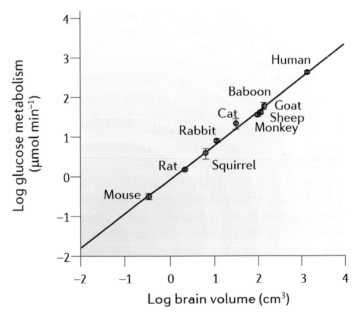

Figure 1.11 Larger brains are metabolically more expensive. These allometric scaling relationships show how important parameters of mammalian brain network organization are constrained by physical size.

Credit: Fig. 1d of ref. 4.

1.5   Summary

The central and peripheral neural system of an animal serves its daily needs, with a variety of modules that have evolved continuously and specifically in each lineage. The number of cells for each function is perhaps more fundamental than the relative proportions of each module. For mammals small and large, the gray matter of the cerebral cortex is 1–3 mm thick, and the number of neuronal cells scales with the surface area. Though alleviated by folding of the surface, the volume of white matter increases more steeply with larger brains and add to the metabolic burden. We'll repeatedly see in later chapters that related functions are close together in the brain, so that there would not be too much extra wiring.

For an average neuron with thousands of synapses on it, the spacing between synapses is in the μm (10^{-6} m) range. Each synapse in the mouse cerebral cortex is about 0.3 μm wide, smaller than a bacterial cell but large enough for proteins[15] (Chapter 5); but synapses appeared larger in the hippocampus[54].

After briefly introducing the sparsely hashed senses in Chapter 2 and development in Chapter 3, this book innovatively depicts a coherent picture of memory and learning that is emerging from recent literature. Chapters 4 and 5 get the readers started on the physical basis of memory from cells to dendritic spines, Chapter 6 discusses the roles of sleeping and dreaming in refreshing the brain and determining what to remember. Chapter 7 completes the list of neuronal cells we have for flexible learning, with how we navigate and store information for space and time. Chapter 8 talks about fascinating topics on mathematics and language, with global coordination, and Chapter 9 approaches adaptive learning and reasoning including counterfactuals, which is a key aspect limiting the utility of AI algorithms since the last century[55]. It also helps to be not too human-centric when we talk about the brain, and the choice of model animals is mostly limited by technology.

Questions

1. What nonmammalian nervous systems do you think we should look into, and in what ways might they be helpful?

2. Besides the metabolic burden, does the increasing number of neuronal cells present problems for the brain as a computational system?

3. When two brain areas have been optimized for different functions and then become close enough for interactions, what new functions may we expect?

References

1. Tosches, M. A. & Laurent, G. Evolution of neuronal identity in the cerebral cortex. *Curr. Opin. Neurobiol.* **56**, 199–208 (2019).

2. Broglio, C. *et al.* Hippocampal Pallium and map-like memories through vertebrate evolution. *J. Behav. Brain Sci.* **05**, 109–120 (2015).

3. Emery, N. J. & Clayton, N. S. Evolution of the avian brain and intelligence. *Curr. Biol.* **15**, R946–50 (2005).

4. Bullmore, E. & Sporns, O. The economy of brain network organization. *Nat. Rev. Neurosci.* **13**, 336–349 (2012).

5. Roth, G. & Dicke, U. Evolution of the brain and Intelligence. *Trends Cogn. Sci.* **9**, 250–257 (2005).

6. Zhang, K. & Sejnowski, T. J. A universal scaling law between gray matter and white matter of cerebral cortex. *Proc. Natl. Acad. Sci. U. S. A.* **97**, 5621–6 (2000).

7. Elston, G. N. *et al.* Specializations of the granular prefrontal cortex of primates: implications for cognitive processing. *Anat. Rec. A. Discov. Mol. Cell. Evol. Biol.* **288**, 26–35 (2006).

8. Shibata, M. *et al.* Hominini-specific regulation of CBLN2 increases prefrontal spinogenesis. *Nature* **598**, 489–494 (2021).

9. Shibata, M. *et al.* Regulation of prefrontal patterning and connectivity by retinoic acid. *Nature* **598**, 483–488 (2021).

10. GF, S. Evolution of the hippocampus in reptiles and birds. *J. Comp. Neurol.* **524**, 496–517 (2016).

11. Pittman, J. T. & Lott, C. S. Startle response memory and hippocampal changes in adult zebrafish pharmacologically-induced to exhibit anxiety/depression-like behaviors. *Physiol. Behav.* **123**, 174–179 (2014).

12. Silva, C. G., Peyre, E. & Nguyen, L. Cell migration promotes dynamic cellular interactions to control cerebral cortex morphogenesis. *Nat. Rev. Neurosci.* **20**, 318–329 (2019).

13. Cadwell, C. R., Bhaduri, A., Mostajo-Radji, M. A., Keefe, M. G. & Nowakowski, T. J. Development and arealization of the cerebral cortex. *Neuron* **103**, 980–1004 (2019).

14. Brodmann, K. *Vergleichende Lokalisationslehre der Grosshirnrinde in ihren Prinzipien dargestellt auf Grund des Zellenbaues/[K. Brodmann]* (1909).

15. Braitenberg, V. & Schüz, A. Cortex: statistics and geometry of neuronal connectivity. *Cortex Stat. Geom. Neuronal Connect* (1998) doi:10.1007/978-3-662-03733-1.

16. Strange, B. A., Witter, M. P., Lein, E. S. & Moser, E. I. Functional organization of the hippocampal longitudinal axis. *Nat. Rev. Neurosci.* **15**, 655–69 (2014).

17. Insausti, R. Comparative anatomy of the entorhinal cortex and hippocampus in mammals. *Hippocampus* **3 Spec No**, 19–26 (1993).

18. Hawrylycz, M. J. *et al.* An anatomically comprehensive atlas of the adult human brain transcriptome. *Nature* **489**, 391–399 (2012).

19. Blumenfeld, H. *Neuroanatomy through Clinical Cases* (Oxford University Press, 2021).

20. Crick, F. C. & Koch, C. What is the function of the claustrum? *Philos. Trans. R. Soc. Lond. B. Biol. Sci.* **360**, 1271–9 (2005).

21. Goll, Y., Atlan, G. & Citri, A. Attention: the claustrum. *Trends Neurosci.* **38**, 486–95 (2015).

22. Peng, H. *et al.* Morphological diversity of single neurons in molecularly defined cell types. *Nature* **598**, 174–181 (2021).

23. Kitanishi, T. & Matsuo, N. Organization of the claustrum-to-entorhinal cortical connection in mice. *J. Neurosci.* **37**, 269–280 (2017).

24. Ardila, A., Bernal, B. & Rosselli, M. How localized are language brain areas? a review of brodmann areas involvement in oral language. *Arch. Clin. Neuropsychol.* **31**, 112–122 (2016).

25. Castellucci, G. A., Kovach, C. K., Howard, M. A., Greenlee, J. D. W. & Long, M. A. A speech planning network for interactive language use. *Nature* 1–6 (2022) doi:10.1038/s41586-021-04270-z.

26. Arcaro, M. J. & Livingstone, M. S. On the relationship between maps and domains in inferotemporal cortex. *Nat. Rev. Neurosci.* (2021) doi:10.1038/s41583-021-00490-4.

27. Carvalho, M. M. *et al.* A Brainstem locomotor circuit drives the activity of speed cells in the medial entorhinal cortex. *Cell Rep.* **32**, 108123 (2020).

28. Lieberman-Aiden, E. *et al.* Comprehensive mapping of long-range interactions reveals folding principles of the human genome. *Science* **326**, 289–93 (2009).

29. Tallinen, T. *et al.* On the growth and form of cortical convolutions. *Nat. Phys.* **12**, 588–593 (2016).

30. Collinet, C. & Lecuit, T. Programmed and self-organized flow of information during morphogenesis. *Nat. Rev. Mol. Cell Biol.* **22**, 245–265 (2021).

31. G. D. *et al.* Perirhinal input to neocortical layer 1 controls learning. *Science* **370** (6523):eaaz3136. doi: 10.1126/science.aaz3136 (2020).

32. Markram, H. *et al.* Interneurons of the neocortical inhibitory system. *Nat. Rev. Neurosci.* **5**, 793–807 (2004).

33. Karimi, A., Odenthal, J., Drawitsch, F., Boergens, K. M. & Helmstaedter, M. Cell-type specific innervation of cortical pyramidal cells at their apical dendrites. *Elife* **9**, e46876 (2020).

34. Lodato, S., Shetty, A. S. & Arlotta, P. Cerebral cortex assembly: generating and reprogramming projection neuron diversity. *Trends Neurosci.* **38**, 117–25 (2015).

35. Saladin, K. S., Gan, C. A. & Cushman, H. N. *Anatomy & Physiology : the Unity of form and Function* (McGraw Hill, 2021).

36. Glassman, R. B. Topology and graph theory applied to cortical anatomy may help explain working memory capacity for three or four simultaneous items. *Brain Res. Bull.* **60**, 25–42 (2003).

37. Charvet, C. J., Cahalane, D. J. & Finlay, B. L. Systematic, cross-cortex variation in neuron numbers in rodents and primates. *Cereb. Cortex* **25**, 147–160 (2015).

38. Ehrlich, B. *The Brain in Search of Itself : Santiago Ramón y Cajal and the Story of the Neuron* (Farrar, Straus and Giroux, 2022).

39. Le Cun, Y. *Quand la machine apprend La révolution des neurones artificiels et de l'apprentissage profond* (Odile Jacob, 2019).

40. Turner, N. L. *et al.* Reconstruction of neocortex: organelles, compartments, cells, circuits, and activity. *Cell* **185**, 1082-1100.e24 (2022).

41. Santuy, A., Rodriguez, J. R., DeFelipe, J. & Merchan-Perez, A. Volume electron microscopy of the distribution of synapses in the neuropil of the juvenile rat somatosensory cortex. *Brain Struct. Funct.* **223**, 77–90 (Basic Books, 2018).

42. Motta, A. *et al.* Dense connectomic reconstruction in layer 4 of the somatosensory cortex. *Science (80-).* **366**: eaay3134. doi: 10.1126/science.aay3134. (2019).

43. Werbos, P. J. Applications of advances in nonlinear sensitivity analysis. in *System Modeling and Optimization* (ed. Drenick, R. F., Kozin, F.) 762–770 (Springer, 1982). doi:10.1007/BFb0006203.

44. LeCun, Y., Bottou, L., Orr, G. B. & Müller, K.-R. Efficient backprop. in *Neuronetworks: Tricks of the Trade* 9–50 (1998). doi:10.1007/3-540-49430-8_2.

45. Dolphin, A. C. & Lee, A. Presynaptic calcium channels: specialized control of synaptic neurotransmitter release. *Nat. Rev. Neurosci.* **21**, 213–229 (2020).

46. Laughlin, S. B., de Ruyter van Steveninck, R. R. & Anderson, J. C. The metabolic cost of neural information. *Nat. Neurosci.* **1**, 36–41 (1998).

47. Karbowski, J. Global and regional brain metabolic scaling and its functional consequences. *BMC Biol.* **5**, 18 (2007).

48. Butterfield, D. A. & Halliwell, B. Oxidative stress, dysfunctional glucose metabolism and Alzheimer disease. *Nat. Rev. Neurosci.* **20**, 148–160 (2019).

49. Mossad, O. *et al.* Gut microbiota drives age-related oxidative stress and mitochondrial damage in microglia via the metabolite N6-carboxymethyllysine. *Nat. Neurosci.* **25**, 295–305 (2022).

50. Ucar, H. *et al.* Mechanical actions of dendritic-spine enlargement on presynaptic exocytosis. *Nature* **600**, 686–689 (2021).

51. Pluvinage, J. V. & Wyss-Coray, T. Systemic factors as mediators of brain homeostasis, ageing and neurodegeneration. *Nat. Rev. Neurosci.* **21**, 93–102 (2020).

52. Olson, C. A. *et al.* The gut microbiota mediates the anti-seizure effects of the ketogenic diet. *Cell* **173**, 1728-1741.e13 (2018).

53. Mattson, M. P., Moehl, K., Ghena, N., Schmaedick, M. & Cheng, A. Intermittent metabolic switching, neuroplasticity and brain health. *Nat. Rev. Neurosci.* **19**, 63–80 (2018).

54. Padmanabhan, P., Kneynsberg, A. & Götz, J. Super-resolution microscopy: a closer look at synaptic dysfunction in Alzheimer disease. *Nat. Rev. Neurosci.* **22**, 723–740 (2021).

55. Pearl, J. & Mackenzie, D. *The Book of Why.* (2018).

Chapter 2

The Senses as Basic Input

Abstract

From marine animals, insects to humans, we are all limited by the resolution of our senses and rewarded through our senses. Each sense detects and locally processes signals. This Chapter illustrates the neuronal setup of basic senses, all of which can be used as cues for navigation (more in Chapter 7) and associative learning. A variety of animals including insects are already inspiring new algorithms and robotics design. Layers of neurons in the visual cortex inspired AI algorithms, but we now know that they work rather differently. It is probably a universal theme that individual functions are achieved with a relatively small number of neurons.

Keywords

Senses, olfaction, sparse network, continual learning, gradient learning, ribbon synapses, vision, data preprocessing, autoencoder, mechanosensory

2.1   Olfaction

2.1.1   Prioritizing with Separation and Tagging?

Olfaction is an ancient sense that plays key roles in both invertebrates (e.g., worms, insects) and vertebrates (those that have a

Neuroscience for Artificial Intelligence
Huijue Jia
Copyright © 2023 Jenny Stanford Publishing Pte. Ltd.
ISBN 978-981-4968-78-2 (Hardcover), 978-1-003-41098-0 (eBook)
www.jennystanford.com

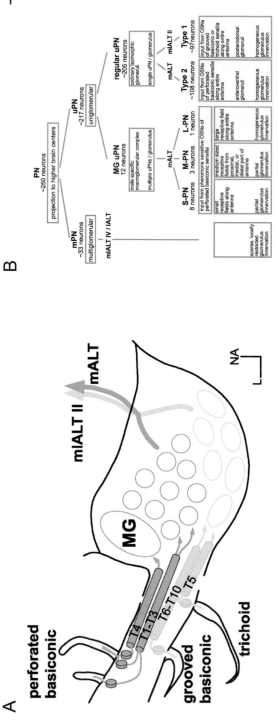

Figure 2.1 Separation of olfactory signals in an insect. (A) Schematic illustration of the olfaction pathways from the periphery to higher brain centers in the cockroach protocerebrum. Olfactory sensory neurons (OSNs) of perforated basiconic sensilla (green) run through antennal tracts T1-T4 and terminate in the n-anteroventral group of glomeruli. Projection neurons (PNs) with dendrites in these glomeruli project through the mALT to the protocerebrum. OSNs in trichoid and grooved basiconic sensilla (yellow) run through antennal tracts T5 and T6-10, respectively, and terminate in the n-posterodorsal group of glomeruli. PNs with dendrites in these glomeruli project through the mALT II to the protocerebrum. L, lateral; mALT, medial antennal lobe tract; MG, macroglomerulus; mlALT II, mediolateral antennal lobe tract; NA, n-anterior; T1-10, antennal tracts 1–10. (B) The number of neurons in each group. Abbreviations as in (A).

Credit: Part A, Fig. 1, and Part B, Fig. 3 of ref. 1.

spine, from fish to humans). In cockroach, olfactory signals, including pheromones, are locally separated and processed in the antenna, before getting into the brain (Fig. 2.1).

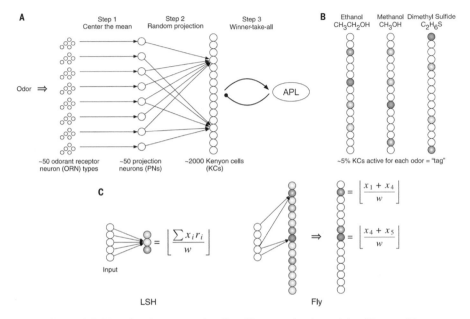

Figure 2.2 Mapping between the fly olfactory circuit and locality-sensitive hashing (LSH). (A) Schematic of the fly olfactory circuit. In step 1, 50 odorant receptor neurons (ORNs) in the fly's nose send axons to 50 projection neurons (PNs) in the glomeruli; as a result of this projection, each odor is represented by an exponential distribution of firing rates, with the same mean for all odors and all odor concentrations. In step 2, the PNs expand the dimensionality, projecting to 2000 Kenyon cells (KCs) connected by a sparse, binary random projection matrix. In step 3, the KCs receive feedback inhibition from the anterior paired lateral (APL) neuron, which leaves only the top 5% of KCs to remain firing spikes for the odor. This 5% corresponds to the tag (hash) for the odor. (B) Illustrative odor responses. Similar pairs of odors (e.g., methanol and ethanol) are assigned more similar tags than are dissimilar odors. Darker shading denotes higher activity. (C) Differences between conventional LSH and the fly algorithm. In the example, the computational complexity for LSH and the fly are the same. The input dimensionality $d = 5$. LSH computes $m = 3$ random projections, each of which requires 10 operations (five multiplications plus five additions). The fly computes $m = 15$ random projections, each of which requires two addition operations. Thus, both require 30 total operations. x, input feature vector; r, Gaussian random variable; w, bin width constant for discretization.

Credit: Fig. 1 of ref. 4.

In general, learning one thing at a time is a prevalent mode in animals, unlike the widely used machine-learning algorithms that try to classify between two or more groups. About 5% of the ~2000 Kenyon cells in a fruit fly are activated by each odor (Fig. 2.2). Architecture of the fruit fly olfactory network (Fig. 2.2) enables lifelong continual learning, with much less catastrophic forgetting than existing continual learning methods using deep neuronal networks such as gradient episodic memory (GEM), elastic weight consolidation (EWC), and brain-inspired replay (BI-R)[2]. The fly olfactory network has been shown to work for natural language processing, with faster training and less memory compared to mainstream algorithms such as global vectors for word representation (GloVe)[3].

2.1.2 The Spatiotemporal Resolution of Olfaction

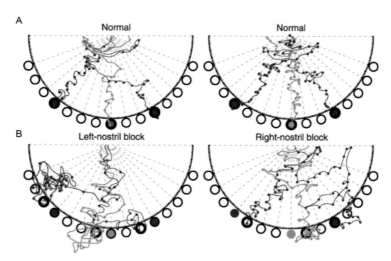

Figure 2.3 Search behavior of moles under normal and nostril-block conditions for radial search paradigm. (A) Normal search behavior from two blind eastern American moles (*Scalopus aquaticus*, animal no. 2 on left and animal no. 1 on right). Color of the trace matches target well color for each trial. Circles indicate sniffs. The first well explored with a nose dip is indicated by bold outline. (B) Examples of search behavior for left- and right-nostril block for cases shown in (A). Left nostril-block biased search to the right, whereas right-nostril block biased search to the left. Colored well indicates target (food) location; bold outline indicates the first well explored. Trace is illustrated until a well was explored, but moles continued to search and found the food in nearly every trial.

Credit: Fig. 2a,b of ref. 10.

Concentration gradients of odors are key information for spatial navigation of animals including humans[5-7] (Chapter 7). Odor concentrations (from air that is ~3.5 cm apart for humans) are compared between nostrils[7]. With one nostril blocked, blind moles would eventually locate the odor source after more sidesteps, and often more towards the unblocked side (Fig. 2.3). Each turn after sniffing is like one step in gradient learning (in AI), with smaller steps closer to the odor source (Fig. 2.3).

Serial sniffing (e.g., 10 times per second, 10 Hz) is not nearly as fast as the natural temporal variations in odor concentration which can reach 1000 Hz with complex aerodynamics, but mice were shown to be able to distinguish between odor frequencies of up to 40 Hz[8] (Fig. 2.4). One possibility is that phasing information relative to a wave (Section 7.8) is utilized for distinguishing between such high frequencies.

We mentioned in Chapter 1 that the forebrain functions primarily in olfaction in fish (Figs. 1.1 and 1.2). The olfactory circuits can detect correlations and anticorrelations (a first step towards causality[9], Chapter 9), and can separate the sources of the odors[8].

2.2   Taste

Taste receptors expressed on the tongue, soft palate and in other places of the body[11], contribute to the reward (or aversion) size perceived by an animal in learning (Chapter 9). Detected at millimolar concentrations, sweets mean calories, and they don't taste as sweet in ice creams. Cats and some bats do not taste sweetness[12]. Bitterness is detected by the same family of receptors that detect sweetness, with some receptors specialized for single poisonous compounds and others more generally activated[11, 13]. Due to genetic variations, phenylthiocarbamide found in vegetables such as brussels sprouts and broccoli tastes bitterer in some people than others[11, 14]. The low pH detected by the sour taste indicates potential toxicity, i.e., spoiled food. Salts can be either nutritious or harmful, depending on the concentration[12]. Some fat that is useful for female fruit flies can be sweet at low concentrations but bitter at high concentrations[15]. In the same type of taste bud cells as sweet and bitter, the Umami taste in humans detects

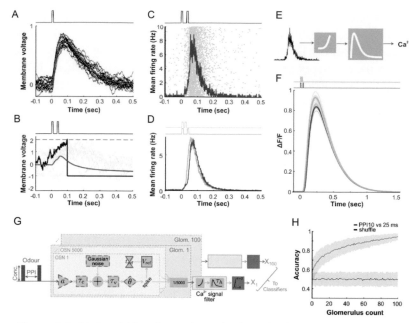

Figure 2.4 Distinguishing fast odor stimuli with slow olfactory sensory neurons (OSNs). (A) Membrane voltage relative to baseline of a single model OSN in response to a 10 ms odor pulse. Black traces are individual trials; red trace is average over 20 trials. OSN spike threshold has been set high enough to prevent spiking to illustrate the subthreshold voltage time course. (B) Membrane voltages (gray traces) of ten OSNs from a population of 5,000 in response to a paired odor pulse with pulse width 10 ms and paired pulse interval (PPI) of 25 ms. The voltage time course for one example OSN is in black. Several OSNs reach the OSN spike threshold (dashed red line) and are temporarily reset to the refractory voltage of −1. The population average membrane voltage (red) reveals membrane charging in response to odor stimulation and the subsequent discharging and refractory period. (C) Raster showing the spike times (dots) of the full population from band the corresponding mean firing rate (trace) estimated in 1 ms bins. (D) Mean firing rates computed over 20 trials in response to paired odor pulses of width 10 ms and PPIs of 10 ms (green) and 25 ms (black). (E) Model calcium signals are produced by squaring the instantaneous mean firing rate and filtering the result with a calcium imaging kernel. (F) Model calcium responses to the paired odor stimulus with a PPI of 10 ms (green) and 25 ms (black). Thin traces are single trials, thick traces are averages over 15 trials. (G) Schematic of the OSN model. Variables in dashed bounding boxes are changed for each glomerulus[8]. (H) Linear classifier analysis over an increasing subset size of glomeruli (1–100; plotted is mean ± s.d., 256 repeats for random subsets of *n* glomeruli generating 256 unshuffled and 256 shuffled accuracies).

Credit: Extended Fig. 1 of ref. 8.

5′-ribonucleotides, glutamate and aspartate (the mice receptor responds to most L-amino acids[12]) which are both a nitrogen source and a carbon source. The Chinese character for Umami is fish and goat. Ancestors of the Giant Panda lost the Umami receptor as they specialized in bamboo[16,17], so it would be maintaining its nitrogen balance in some other way. Pungency is detected by heat-activated cation channels related to pain sensation[11]. Astringency, a dry, rough or puckering mouthfeel, is likely due to tactile sensation of aggregation of lubricant proteins in the saliva (due to chemicals in wine, tea, unripe fruit, etc.)[18].

2.3   Hearing

Unlike visual scenes in the environment, sounds always mean that someone or something has made the sound[19]. If a group of animals are calling, one has to know what to do. We can attest to the distracting nature of unattended auditory information that is also processed. In primates like us, a lot of spectral details are already lost at the cochlea in the inner ear, with a roughly logarithmic representation of increasing frequency of sound[19]. Mechanotransduction at the bundled ciliates of hair cells rapidly generates changing receptor potentials, but this changing signal is limited by the membrane time constant. Instead of the classic point-to-point transduction at synapses (Chapter 1), hair cells in the inner ear, as well as rods and cones in the retina, have ribbon synapses that enable persistent activity[20] (Fig. 2.5).

Differences in sound intensity, time delay and phase between the two ears can all be useful for locating the direction of the sound[21] (Fig. 2.6). The spatial resolution in the horizontal plane is highest around the interaural midline in front of an individual and deteriorates towards the acoustic periphery, especially behind one's back. The posterior auditory cortex responds faster, while the anterior auditory cortex has a more sustained activity[19, 21] (Fig. 2.6).

Auditory hallucinations are common for patients with conditions such as schizophrenia and bipolar disorders (Fig. 2.7), whereas Parkinson's disease patients treated with L-dopa could have visual and auditory hallucinations[22, 23]. Injuries in the auditory cortex did not result in deafness, but the patients became "word deaf" and can no longer process the structure information in sound (more on language in Chapter 8).

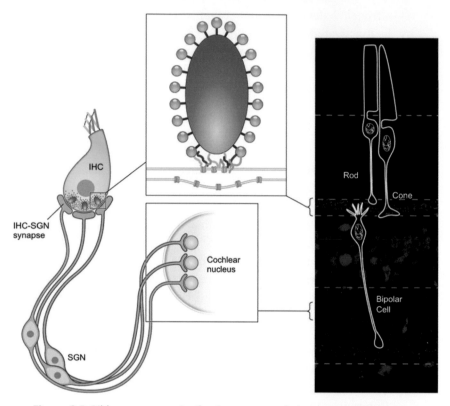

Figure 2.5 Ribbon synapses in the inner ear and the retina. With one or more "ribbon" anchored presynaptically, glutamate can be persistently released in the active zone. Left: Signals from inner hair cells (IHC) go through ribbon synapses with spiral ganglion cells (SGN). Deflection of hair bundles activates mechanosensitive channels on the hair cells of the auditory and vestibular epithelia of the inner ear, which then release multiple doses of glutamate to the SGN. Right: Photon-induced isomerization of retinal hyperpolarize rods and cones, which then signal to bipolar cells through ribbon synapses. Bipolar cells again have ribbon synapses to signal to amacrine and ganglion cells in the inner retina.

Credit: From Abstract of ref. 20.

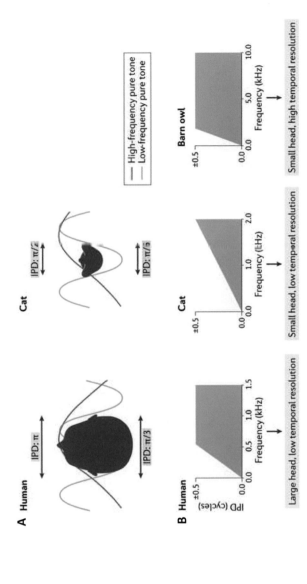

Figure 2.6 The physiological range of interaural phase differences (IPD) is a function of sound frequency, head size and temporal resolution of the auditory system. For periodic tones, interaural time differences (ITD) up to half the period of the tone can be expressed in terms of phase shifts, i.e., IPD. (A) Detection of IPD depends on head size. Humans have a relatively large head size and IPD is more useful than ILD (interaural level (intensity) difference) only for low frequencies, which include the frequency of speech. The legend for high-frequency and low-frequency pure tones have been swapped from ref. 21. (B) Besides head size, IPD is also limited by the temporal resolution of the auditory system, i.e., the highest frequency to which the firing of auditory nerve fibers and neurons at subsequent stages in the subcortical auditory pathway can phase lock, which reaches 10 kHz in barn owls.

Credit: From Box 1 of ref. 21.

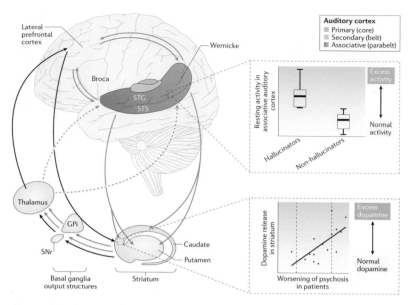

Figure 2.7 Circuitry of associative auditory cortex and basal ganglia relevant to perceptual disturbances in psychosis. The anatomy of auditory cortex and the downstream projections of associative auditory cortex to prefrontal cortex and striatum, as well as the cortico-cortical loops and cortico-basal ganglia-thalamocortical loops proposed to be relevant for perceptual disturbances. In the center, a lateral surface of the brain is shown, indicating the different areas of auditory cortex in different colors. The associative auditory cortex features a gradient from anterior-ventral to posterior-dorsal aspects. Connections from these aspects of the associative auditory cortex to the dorsal striatum and lateral prefrontal cortex are indicated by arrows reflecting the colors of the areas of origin. Note that the striatum, basal ganglia and thalamus output structures are in the center of the brain, but here they are shown separately for illustrative purposes. Prefrontal inputs are also shown as gray arrows. Outputs from the striatum to basal ganglia output structures and the thalamus are also illustrated, in addition to thalamocortical projections back to the areas where the inputs originate. Dashed arrows indicate connections that have not been fully described in primates. Findings of increased activity in associative auditory cortex in relation to hallucinations and increased striatal dopamine in psychosis are depicted as inserts.

Credit: From Fig. 4a of ref. 23.

2.4 Visual Signal Processing in Each Cell

Vision is a dominant sense in many animals. Among the rods and cone photoreceptor cells in the retina (Fig. 2.5), the cones detect colors, and motion blur that can be decoded for movements[24].

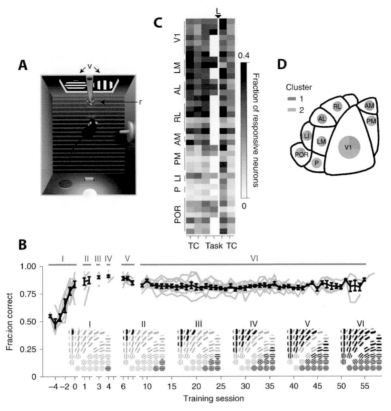

Figure 2.8 Mice learning two visual categories with neuronal activities in the visual cortex. (A) Mice learn discriminating information-integration categories in a touch screen operant chamber. Touch screens (v) display visual stimuli and record screen presses, two at a time from a total of 42 grating stimuli that differed in orientation and spatial frequency. Food pellet rewards (r) are delivered via a pellet feeder into a dish, after each correct choice of the rewarded category, and not the other category. A drinking bottle, house light, speaker, and lever are positioned on the east and south walls. (B) Mean (±s.e.m.) learning curve. Gray lines represent individual animals (n = 8 mice). Latin numerals denote category training stages. Insets show active stimuli (black) and not-yet-introduced stimuli (gray). (C,D) With head fixed for chronic calcium imaging, unlike in (A), only one visual stimulus was shown each time, and the mouse's choice was indicated by licking for a water reward. (C) Fraction of stimulus- and task-activated responsive neurons (corrected for variable trial numbers by subsampling). The y axis shows visual areas with chronic recordings, and the x axis shows imaging time points. TC, out-of-task time points (tuning curves). Task, in-task time points. L, separates baseline from after category learning. The second in-task baseline was only acquired in a subset of mice. (D) Map of mouse

(Continued)

Figure 2.8 (*Continued*)

visual areas (based on ref. 25) showing the fraction of cluster 1 and cluster 2 (k-means clustering) chronic recordings per area after category learning[26]. V1, primary visual cortex; LM, lateromedial area; LI, laterointermediate area; AL, anterolateral area; RL, rostrolateral area; AM, anteromedial area; PM, posteromedial area; P, posterior; POR, postrhinal. The mouse V1 has direct input to other visual areas, while only V2, V3, V4 and the middle temporal area (MT) are known to receive substantial V1 input in the primate brain[25].

Credit: Part A,B from Fig. 1a,b and Part C,D from Fig. 4a,d of ref. 26.

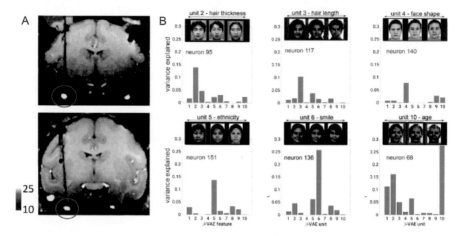

Figure 2.9 Responses of single neurons are well explained by single disentangled latent units. (A) Coronal section showing the location of fMRI-identified face patches in two primates, with patch AM (anteromedial area) circled in red. Dark black lines, electrodes. (B) Explained variance of single neuron responses to 2,100 faces. Response variance in single neurons is explained primarily by single disentangled units encoding different semantically meaningful information, using a deep unsupervised generative model, β-variational autoencoder (beta-VAE). Insets, latent traversals were used to visualize the semantic meaning encoded by single disentangled latent units of a trained model.

Credit: Fig. 2 of ref. 27.

Different areas of the visual cortex (with hundreds to low thousands of neurons in each area in mice) already differ in sensitivity for different features, and there are crosstalk between areas[25, 28, 29]. Mice readily learn categories of visual stimuli in the first few trials, using a fraction of the neurons (Fig. 2.8). Intuitively, the category with a food or water reward (or something fearful) might be more important (Fig. 2.8).

Layer I (L1) of the neocortex mostly contained dendrites from the deeper layers (L2, L3, L5 and occasionally L4[30-32], Fig. 1.9). L1 and L4 of the primary visual cortex (V1) receives input from, and L5-L6 output to evolutionarily more ancient regions such as the thalamus, which includes information on locomotion[28, 33] (and the thalamus is important both when awake and while asleep, Chapter 6).

Spontaneous activities in the visual cortex start before eye-opening. Synchronous firing occurs in adjacent neurons sharing similar orientation or spatial frequency preferences in L2-L4 (including excitatory interneurons, i.e., stellate cells, Fig. 1.9) of the visual cortex in rats after eye-opening, whereas neurons in the deeper layers L5-L6 did not get so synchronous with the stimulus[34] (and L6 output to the claustrum[35], Section 7.8). Consistently, adult mice showed more connections within the same layer in the superficial layers than in the deeper layers[36].

Single neurons in the visual cortex already encode distinct features of an image, such as the age, smile, hair thickness of a face (Fig. 2.9). We do tend to see smiling faces in objects (searching and hashing, Section 4.4). The face is part of a two-dimensional (2D) continuum of objects recognized: animate versus inanimate in one dimension, stubby (round) versus spiky in the other dimension (Fig. 2.10, more speculations on written language in Section 8.7). Such an arrangement is likely both genetic and stimulus-modulated[37]. Puppies got to see human faces no later than they open their eyes and see their mother, and learned to read our facial expressions. Animals that look alike to us can look rather unique to each other. Fish[38] and birds have many neighbors. Crayfish have been shown to recognize a fight opponent, according to spots and head width[39].

Objects can also be reconstructed from signals recorded in single neurons, together with a database of labelled images (Fig. 2.11; our brain is also probably generative according to a database, Section 7.8). Thus, if there are scratches on one's photo (Fig. 2.12), our brain probably detects the scratches and the face with separate cells, unlike the statistical averaging approaches developed by data scientists.

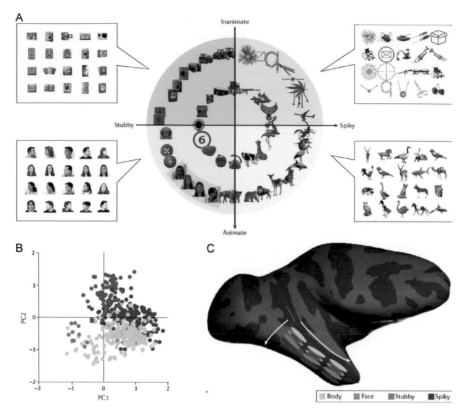

Figure 2.10 The neural code for object identity in general. (A) Schematic showing the four quadrants spanned by the first two principal components (PCs) of object space, animate–inanimate (PC1) and stubby–spiky (PC2). The stimuli in each of the quadrants were used for mapping four networks (face patches, body patches, spiky patches and stubby patches) using functional MRI. (B) Projection of preferred axis of each cell (n = 482) onto PC1 versus PC2 for all neurons recorded across four networks (spiky network: orange, body network: lime, face network: green, stubby network: blue). (C) Schematic showing the threefold-repeated topographic map in the monkey inferotemporal cortex that is organized according to the four quadrants of object space. In humans, the fusiform gyrus (Brodmann Area 37) appeared more rolled into the bottom of the lobe (e.g., ref. 40), reminiscent of the different configuration of the hippocampus (Fig. 1.6).

Credit: Fig. 6a,b,c of ref. 41, (A) and (B) were adapted by ref. 41 from ref. 42.

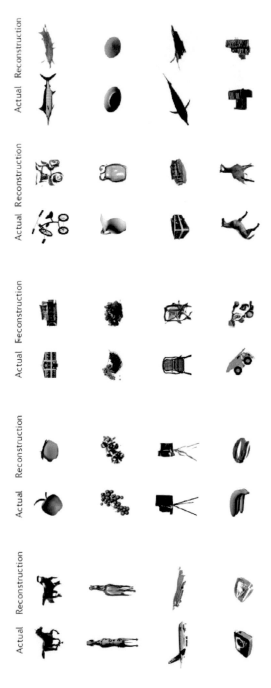

Figure 2.11 Random objects were reconstructed using the activity of 482 cells sampled from the four networks in Fig. 2.10 (face patches, body patches, spiky patches and stubby patches) and a generative adversarial network[42]. The authors passed an image set containing 18,700 background-free object images (http://www.freepngs.com) and 600 face images (FEI database), none of which had been shown to the monkey, through AlexNet, and projected these images to the object space computed using the original stimulus set of 1,224 images. For each image, the object feature vector reconstructed from neural activity was compared with object feature vectors for images from the new image set. The image in the new image set with the smallest Euclidean distance to the reconstructed object feature vector was considered as the "reconstruction" of this object feature vector.

Credit: Fig. 6d of ref. 41, adapted by ref. 42.

Figure 2.12 A photo with scratches.

Credit: Huijue Jia.

2.5 Sensing Mechanical Forces

Mechanical forces, whether dangerous or appealing, have to be properly sensed. Insects have cells that reach into leg joints to convert displacements into neuronal signals[43]. Mechanosensory cells in the nonhairy glabrous skin in mammalian hands and feet develop after birth to properly project to many neurons in the cortex (Fig. 1.7) and the brainstem[44, 45].

Neurons in the mouse barrel cortex correspond to individual whiskers, which have a high spatiotemporal sensitivity (~5 ms, Fig. 2.13) but can be tuned down during active whiskering[46]. The whiskers can sense shallow grooves < 100 μm apart, similar to the texture resolution of human finger prints[46]. A stronger input could possibly involve more layers in the barrel cortex (Fig. 2.14) and spread in ~20 ms, whereas a single activated axon would turn off in 5–50 ms[46].

Hairy skin contains distinct types of neurons for hair with different distributions (Fig. 2.15), which help distinguish among various types and sources of touch, e.g., a gentle stroke[44, 47].

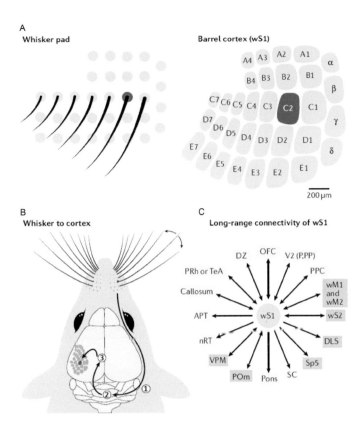

Figure 2.13 Organization and connectivity of the whisker-related primary somatosensory cortex. (A) The whisker-related primary somatosensory cortex (wS1) of rats and mice contains obvious anatomical units called "barrels", each of which represents an individual mystacial whisker and is somatotopically organized. (B) Deflection of a mystacial whisker evokes a sequence of activity in trigeminal ganglion primary sensory neurons (1), brainstem neurons (2) and thalamic neurons (3) before reaching wS1. (C) A schematic representation of the long-range connectivity of wS1. Red boxes highlight strongly connected brain regions discussed further in this Review. APT, anterior pretectal nucleus; DLS, dorsolateral striatum; DZ, dysgranular zone surrounding wS1; nRT, nucleus reticularis of the thalamus; OFC, orbitofrontal cortex; POm, posterior medial nucleus of the thalamus; PPC, posterior parietal cortex; PRh, perirhinal cortex; SC, superior colliculus; Sp5, spinal trigeminal nuclei; TeA, temporal association cortex; wM1, whisker-related primary motor cortex; wM2, whisker-related secondary motor cortex; wS2, whisker-related secondary somatosensory cortex; VPM, ventral posterior medial nucleus of the thalamus; V2 (P,PP), secondary visual area (labelled in previous studies as area P or PP).

Credit: Fig. 1 of ref. 46.

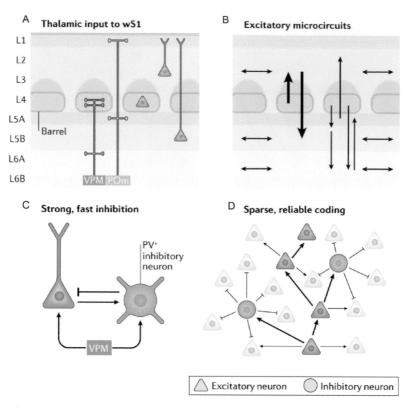

Figure 2.14 Neural circuits for sparse reliable coding of touch in the whisker-related primary somatosensory cortex. (A) Neurons in the whisker-related primary somatosensory cortex (wS1) receive input from the ventral posterior medial nucleus of the thalamus (VPM), which signals whisker deflections primarily to the layer 4 (L4) barrels (blue outlines). VPM axons (blue shading) extend into the L3 regions directly above the L4 barrels. Higher-order thalamic input from the posterior medial nucleus of the thalamus (POm, green) innervates L1 and L5A. (B) Excitatory neuronal microcircuits of wS1 include the "canonical" L4→L2/3→L5 pathway, as well as many other synaptic pathways, including L4→L5, L4→L6, L5A→L2, L6→L5A and L5→L6 (vertical arrows). Extensive horizontal connectivity (horizontal arrows) across barrel columns is prominent in L2/3 and L5/6. (C) Fast-spiking inhibitory GABAergic neurons that express parvalbumin (PV) are strongly and reciprocally connected to nearby excitatory neurons. Both inhibitory and excitatory neurons also receive thalamic input. PV+ neurons provide feed-forward, lateral and feedback inhibition. (D) Sparse strong excitatory synaptic connectivity combined with strong dense inhibition could drive reliable, sparse activity in specifically wired excitatory neuronal circuits.

Credit: Fig. 2 of ref. 46.

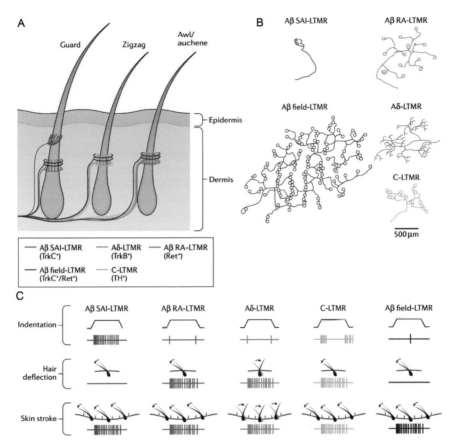

Figure 2.15 The mechanoreceptors of mammalian hairy skin. (A) Blanketing most of the body of mice and most mammals are several types of hairs (in mice, they are the guard, zigzag and awl/auchene hairs) that receive unique combinations of innervation by distinct low-threshold mechanoreceptor (LTMR) subsets. In mice, while the Ret-expressing (Ret+) Aβ rapidly adapting (RA) LTMRs, TrkB-expressing (TrkB+) Aδ-LTMRs and TH-expressing (TH+) C-LTMRs all form lanceolate endings, the TrkC+/Ret+ Aβ field-LTMRs form circumferential endings and the TrkC-expressing (TrkC+) Aβ slowly adapting type 1 (SAI) LTMRs form touch domes with Merkel cells (shown in green). In mice (as shown here), the Aβ SAI-LTMRs form touch domes restricted to the mouth of the guard hair (the longest but least prevalent body hair); however, in humans, the Aβ SAI-LTMRs/touch domes are typically localized to skin between hair follicles. (B) A comparison of the peripheral terminals of individual, sparsely labelled LTMR subtypes in trunk hairy skin reveals that the lanceolate ending-forming and circumferential ending-forming LTMRs innervate numerous hair follicles (sometimes more than 100 hair

(*Continued*)

Figure 2.15 (*Continued*)

follicles in the case of the Aβ field-LTMR) over a large terminal area, unlike the Aβ SAI-LTMRs, which have a restricted terminal field area within the skin (usually associated with only a single guard hair in the mouse). The Aδ-LTMRs are unique in having polarized lanceolate endings around hair follicles that are restricted to the caudal (downward) side of zigzag and awl/auchene follicles. (C) Each LTMR subtype innervating hairy skin has a unique set of responses to common tactile stimuli. The responses shown here are schematic illustrations of typical responses. Although Aβ SAI-LTMRs, Aβ RA-LTMRs, Aδ-LTMRs, C-LTMRs and Aβ field-LTMRs all respond to a stroke, which indents the skin and deflects multiple hairs as the stimulus moves across the surface of the skin, the Aβ SAI-LTMRs and Aβ field-LTMRs are insensitive to deflection of individual hairs, and the Aδ-LTMRs are sensitive to deflection of awl/auchene or zigzag hairs in trunk skin specifically along the caudal-to-rostral axis and to deflection of the hairs in ventral paw skin along the rostral-to-caudal axis. The Aβ field-LTMRs, while highly sensitive to gentle stroking across the skin, are relatively insensitive to skin indentation and thus exhibit indentation responses characteristic of high-threshold mechanoreceptors[47]. Although the Aβ SAI-LTMR, Aβ RA-LTMR, Aδ-LTMR and C-LTMRs all respond to skin indentation at low forces, their rates of adaptation vary, with the Aβ RA-LTMRs and Aδ-LTMRs adapting the fastest and the Aβ SAI-LTMRs the slowest. The Aβ field-LTMRs adapt rapidly at the lowest indentation forces for which they do respond, while at high forces, the responses of the Aβ field-LTMRs to indentation steps are more intermediately adapting and lack an off response[47].

Credit: Fig. 2 of ref. 44. Part B was adapted by ref. 44 with permission from ref. 47.

2.6 Summary

From marine animals, insects to humans, we are all limited and shielded by the spatiotemporal resolution of our senses, which underlie what we remember and respond to. As long as an organism gets the necessary information for what it needs to do (e.g., bacterial chemotaxis[48]), more information is not necessarily better. It is probably a universal theme that individual functions are achieved with a relatively small number of neurons. The sensory input sparsely and locally activates some of the neurons. Different inputs can be correlated and tracked. At the sensory end, the neural system is likely already a bottom-up setup that detects useful information (even when asleep, Chapter 6), while it can be modulated by a general status communicated as an integrated message.

Questions

1. How many steps do you think are needed from input to function for each sense?

2. When a group of cells become more sensitive to a particular input, what would be a good distribution of very sensitive cells and slightly more sensitive cells? How easily should they be reallocated to be more sensitive to another input?

3. What other information from the environment could be useful for an animal, and how might that be sensed?

References

1. Fuscà, D. & Kloppenburg, P. Odor processing in the cockroach antennal lobe-the network components. *Cell Tissue Res* **383**, 59–73 (2021).

2. Shen, Y., Dasgupta, S. & Navlakha, S. Algorithmic insights on continual learning from fruit flies (2021). https://doi.org/10.48550/arXiv.2107.07617.

3. Liang, Y. *et al.* Can a Fruit Fly Learn Word Embeddings? (2021). https://doi.org/10.48550/arXiv.2101.06887.

4. Dasgupta, S., Stevens, C. F. & Navlakha, S. A neural algorithm for a fundamental computing problem. *Science (80-.).* **358**, 793–796 (2017).

5. DH, G., V, K., A, A.-A., A, S. & VN, M. Mice develop efficient strategies for foraging and navigation using complex natural stimuli. *Curr. Biol.* **26**, 1261–1273 (2016).

6. Bao, X. *et al.* Grid-like neural representations support olfactory navigation of a two-dimensional odor space. *Neuron* **102**, 1066–1075. e5 (2019).

7. Wu, Y., Chen, K., Ye, Y., Zhang, T. & Zhou, W. Humans navigate with stereo olfaction. *Proc. Natl. Acad. Sci.* **117**, 16065–16071 (2020).

8. Ackels, T. *et al.* Fast odour dynamics are encoded in the olfactory system and guide behaviour. *Nature* **593**, 558–563 (2021).

9. Pearl, J. & Mackenzie, D. *The Book of Why* (Basic Books, 2018).

10. KC, C. Stereo and serial sniffing guide navigation to an odour source in a mammal. *Nat. Commun.* **4** (2013).

11. Roper, S. D. & Chaudhari, N. Taste buds: cells, signals and synapses. *Nat. Rev. Neurosci.* **18**, 485–497 (2017).

12. Demi, L. M., Taylor, B. W., Reading, B. J., Tordoff, M. G. & Dunn, R. R. Understanding the evolution of nutritive taste in animals: insights from biological stoichiometry and nutritional geometry. *Ecol. Evol.* **11**, 8441–8455 (2021).

13. Yarmolinsky, D. A., Zuker, C. S. & Ryba, N. J. P. Common Sense about Taste: from Mammals to Insects. *Cell* **139**, 234–244 (2009).

14. Dotson, C. D., Shaw, H. L., Mitchell, B. D., Munger, S. D. & Steinle, N. I. Variation in the gene TAS2R38 is associated with the eating behavior disinhibition in Old Order Amish women. *Appetite* **54**, 93–99 (2010).

15. Ahn, J. E., Chen, Y. & Amrein, H. Molecular basis of fatty acid taste in Drosophila. *Elife* **6** (2017).

16. Li, R. *et al.* The sequence and de novo assembly of the giant panda genome. *Nature* **463**, 311–7 (2010).

17. Feng, P. & Zhao, H. Bin. Complex evolutionary history of the vertebrate sweet/umami taste receptor genes. *Chinese Sci. Bull.* **58**, 2198–2204 (2013).

18. Kim, M., Heo, G. & Kim, S.-Y. Neural signalling of gut mechanosensation in ingestive and digestive processes. *Nat. Rev. Neurosci.* **23**, 135–156 (2022).

19. Jasmin, K., Lima, C. F. & Scott, S. K. Understanding rostral–caudal auditory cortex contributions to auditory perception. *Nat. Rev. Neurosci.* **20**, 425–434 (2019).

20. Moser, T., Grabner, C. P. & Schmitz, F. Sensory processing at ribbon synapses in the retina and the cochlea. *Physiol. Rev.* **100**, 103–144 (2020).

21. van der Heijden, K., Rauschecker, J. P., de Gelder, B. & Formisano, E. Cortical mechanisms of spatial hearing. *Nat. Rev. Neurosci.* **20**, 609–623 (2019).

22. Llorca, P. M. *et al.* Hallucinations in schizophrenia and Parkinson's disease: an analysis of sensory modalities involved and the repercussion on patients. *Sci. Rep.* **6**, 38152 (2016).

23. Horga, G. & Abi-Dargham, A. An integrative framework for perceptual disturbances in psychosis. *Nat. Rev. Neurosci.* **20**, 763–778 (2019).

24. Silverstein, S. M. & Rosen, R. Schizophrenia and the eye. *Schizophr. Res. Cogn.* **2**, 46–55 (2015).

25. Marshel, J. H., Garrett, M. E., Nauhaus, I. & Callaway, E. M. Functional specialization of seven mouse visual cortical areas. *Neuron* **72**, 1040–1054 (2011).

26. Goltstein, P. M., Reinert, S., Bonhoeffer, T. & Hübener, M. Mouse visual cortex areas represent perceptual and semantic features of learned visual categories. *Nat. Neurosci.* **24**, 1441–1451 (2021).

27. Higgins, I. *et al.* Unsupervised deep learning identifies semantic disentanglement in single inferotemporal face patch neurons. *Nat. Commun.* **12**, 6456 (2021).

28. Bennett, C. *et al.* Higher-order thalamic circuits channel parallel streams of visual information in mice. *Neuron* **102**, 477-492.e5 (2019).

29. D'Souza, R. D. *et al.* Hierarchical and nonhierarchical features of the mouse visual cortical network. *Nat. Commun.* **13**, 503 (2022).

30. G, D. *et al.* Perirhinal input to neocortical layer 1 controls learning. *Science* **370** (2020).

31. Peng, H. *et al.* Morphological diversity of single neurons in molecularly defined cell types. *Nature* **598**, 174–181 (2021).

32. Karimi, A., Odenthal, J., Drawitsch, F., Boergens, K. M. & Helmstaedter, M. Cell-type specific innervation of cortical pyramidal cells at their apical dendrites. *Elife* **9** (2020).

33. Roth, M. M. *et al.* Thalamic nuclei convey diverse contextual information to layer 1 of visual cortex. *Nat. Neurosci.* **19**, 299–307 (2016).

34. AW, I., Y, K. & Y, Y. Experience-Dependent Development of Feature-Selective Synchronization in the Primary Visual Cortex. *J. Neurosci.* **38**, 7852–7869 (2018).

35. Goll, Y., Atlan, G. & Citri, A. Attention: the claustrum. *Trends Neurosci.* **38**, 486–95 (2015).

36. Campagnola, L. *et al.* Local connectivity and synaptic dynamics in mouse and human neocortex. *Science (80-.).* **375** (2022).

37. Arcaro, M. J. & Livingstone, M. S. On the relationship between maps and domains in inferotemporal cortex. *Nat. Rev. Neurosci.* **22**, 573–583 (2021).

38. Balcombe, J. *What a Fish Knows: The Inner Lives of Our Underwater Cousins* (Scientific American, 2017).

39. Van der Velden, J., Zheng, Y., Patullo, B. W. & Macmillan, D. L. Crayfish recognize the faces of fight opponents. *PLoS One* **3**, e1695 (2008).

40. Norman, Y. *et al.* Hippocampal sharp-wave ripples linked to visual episodic recollection in humans. *Science* **365** (2019).

41. Hesse, J. K. & Tsao, D. Y. The macaque face patch system: a turtle's underbelly for the brain. *Nat. Rev. Neurosci.* **21**, 695–716 (2020).

42. Bao, P., She, L., McGill, M. & Tsao, D. Y. A map of object space in primate inferotemporal cortex. *Nature* **583**, 103–108 (2020).

43. Harris, C. M. *et al.* Gradients in mechanotransduction of force and body weight in insects. *Arthropod Struct. Dev.* **58**, 100970 (2020).

44. Handler, A. & Ginty, D. D. The mechanosensory neurons of touch and their mechanisms of activation. *Nat. Rev. Neurosci.* **22**, 521–537 (2021).

45. Lehnert, B. P. *et al.* Mechanoreceptor synapses in the brainstem shape the central representation of touch. *Cell* **184**, 5608-5621.e18 (2021).

46. Petersen, C. C. H. Sensorimotor processing in the rodent barrel cortex. *Nat. Rev. Neurosci.* **20**, 533–546 (2019).

47. Bai, L. *et al.* Genetic identification of an expansive mechanoreceptor sensitive to skin stroking. *Cell* **163**, 1783–1795 (2015).

48. Mattingly, H. H., Kamino, K., Machta, B. B. & Emonet, T. *Escherichia coli* chemotaxis is information limited. *Nat. Phys.* **17**, 1426–1431 (2021).

Chapter 3

Changing Priorities with Age

Abstract

Young animals learn to sense the environment and to move around, while kept relatively safe. Their neural system grows in the process and matures not much later after they reach adult size. Despite individual differences, the network of different modules in the brain matures in childhood and adolescence. Older people do not sleep as much, and their memories might be kept as an evolutionary treasure that can be key to the survival of a population in difficult years.

Keywords

Cerebellum development, autism spectrum disorder (ASD), Feedforward bias, modularization, default network, Anterior hippocampus, Alzheimer's disease, Sleep fragmentation

3.1 Growing and Learning with the Cerebellum

We are born confused, even with eyes open. As everything is still growing, a challenging task for the neural system is not to be too awkward (and get caught by a predator), but also not too perfect that the connections would have to be revamped for adult dimensions. For example, as the head size of an owl grows, the auditory circuits grow as well, and mature only after reaching the adult size[1]. The vision develops with spontaneous activities and

Neuroscience for Artificial Intelligence
Huijue Jia
Copyright © 2023 Jenny Stanford Publishing Pte. Ltd.
ISBN 978-981-4968-78-2 (Hardcover), 978-1-003-41098-0 (eBook)
www.jennystanford.com

with stimuli (Chapter 2). A pleasant smell could be remembered for decades.

The cerebellum is part of the hindbrain (Chapter 1, Figs. 1.1 and 1.5) that coordinates movements and is implicated in many functions. The new neurons migrate to their proper places before and after birth, and gradually interconnect into an efficient network (Figs. 3.1 and 3.2). Purkinje cells form elaborate branches and also have dendritic spines. The final output is inhibitory (Fig. 3.2).

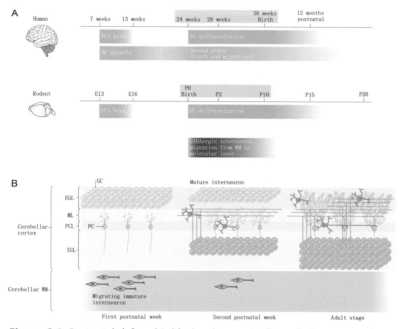

Figure 3.1 Postnatal (after birth) development of cerebellar connections. (A) When the human and rodent timelines are aligned on the basis of major cellular and/or developmental events in the cerebellum, in humans the window of vulnerability to injury (indicated by pink shading) is mostly late gestational, whereas in preclinical rodent models it is mostly postnatal. (B) Cellular schematic of events depicted in the timeline in panel (A) showing external granule layer (EGL) expansion (light green), dendritic arborization of PCs (blue) and white matter (WM) interneuron migration in the first postnatal week. Migration of GCs into the internal granule layer (IGL; dark green) continues in the second postnatal week with concomitant reduction in EGL and circuit formation. In the adult, formation of cerebellar circuitry is completed, the EGL has disappeared and migrating immature interneurons have been integrated into the cerebellar cortical circuitry. A, anterior; D, dorsal; E, embryonic day; GL, granule layer; L, lateral; M, medial; ML, molecular layer; P, posterior; P0, postnatal day 0; PCL, Purkinje cell layer; V, ventral.

Credit: Fig. 1c, d of ref. 3. Part A was adapted by ref. 3 from ref. 10.

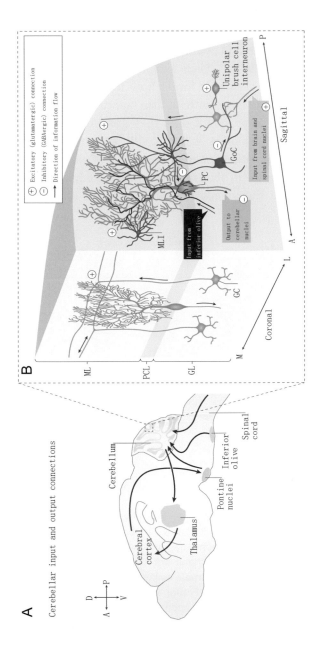

Figure 3.2 Essential features of cerebellar connections and circuitry. (A) General scheme of input and output connections to and from the cerebellum. Main inputs include the spinal cord, inferior olive and pontine nuclei. Main outputs include connections from cerebellar nuclei to the cerebral cortex via the thalamus. (B) Cellular anatomy and circuit connections within the cerebellar cortex. Purkinje cells (PCs) are shown in light blue, granule cells (GCs) in green, molecular layer interneurons (MLIs) in salmon pink, Golgi cells (GoCs) in dark green, unipolar brush cell interneuron in purple, input from inferior olive is shown in dark blue and input from the brain and spinal cord is shown in gray.

Credit: Fig. 1a, b of ref. 3.

A typical pyramidal neuron in the mouse neocortex has approximately 8,000 synapses[2]. In contrast, it is estimated that a single Purkinje cell (PC) in the cerebellum may have around 200,000 synapses[3]. Parallel fibers from granule cells usually make a single synapse with a PC, while climbing fibers make over 500 synapses with a single PC[4]. Such a capacity and robustness may be one reason why we never forget how to ride a bike or how to rope-skip. While those with autism spectrum disorder (ASD) could appear not as awkward once they grow up, adaptive learning through the cerebellum is often still not as easy for those with ASD as for other people (Fig. 3.3). Cells in the cerebellum also participate in the integration and modulation of multiple senses, which then feeds information to the thalamus (Fig. 3.4)[3].

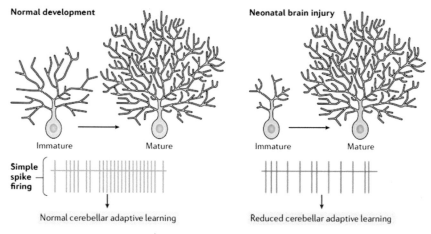

Figure 3.3 In a mouse model of neonatal brain injury, which is a risk factor for ASD, PC dendritic formation is delayed and spike patterns are altered. Delayed development of PC dendritic arborizations during the first 2 postnatal weeks likely causes alterations in PC circuitry and physiology, resulting in significant long-term learning deficits. Reduced numbers of synapses are a characteristic of many neurodevelopmental diseases. Mutations in the postsynaptic scaffolding protein family shank are strongly associated with ASD. PCs in Shank2-knockout mice fire simple spikes with altered regularity and display synaptic plasticity deficits. Future experiments may help to determine specific synaptic PC deficits in neonatal brain injury models to address potential convergence with knockout and mutant models.

Credit: Box 1b of ref. 3.

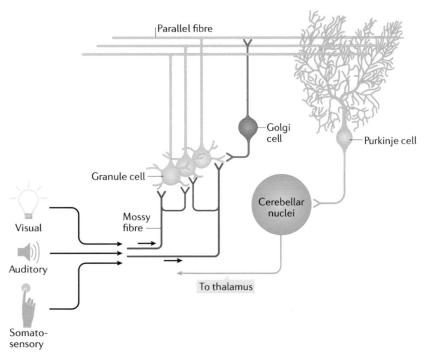

Figure 3.4 Multisensory integration in the cerebellum. Granule cells (light green) integrate sensory information (auditory, visual and somatosensory), resulting in multisensory integration. Golgi cells (green) are also involved in the modulation of sensory information flowing into the cerebellum. Integration of sensory information from complex stimuli such as speech is impaired in autism spectrum disorder.

Credit: Fig. 3b of ref. 3.

Besides, some mutations that often occur in ASD could skew the development of cortical neurons, especially γ-aminobutyric-acid-releasing (GABAergic) interneurons and deep-layer excitatory projection neurons[5], which might then impact sleep (Chapter 6), cognition (Chapter 9) and the sensing of speed (Chapter 7). In addition to dedicated sources of interneurons during development, recent studies have found cells in the cortex and in the cerebellum that can give rise to both excitatory and inhibitory neurons[6–8]; Pyramidal neurons regulate the survival of interneurons as they develop proper ratios and connections[9].

3.2 A Cortical Network that Ripens with Age

For most animals, the number of synapses increases after birth, as the cells grow and contact one another. In the worm *Caenorhabditis elegans*, neurites grow uniformly according to the shapes and relative positions that have been laid out at birth[11]. As the worm matures into adulthood, however, the connections from sensory input to action become increasingly feedforward and modular[11] (Fig. 3.5). So the brain becomes more reflexive with age.

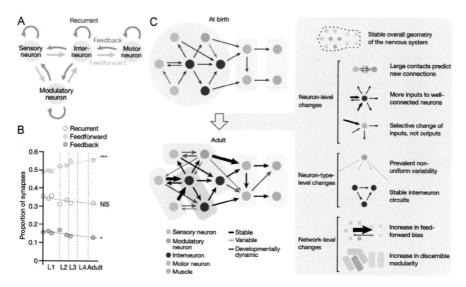

Figure 3.5 Developmental increase in feedforward signaling and modularity in the *Caenorhabditis elegans* brain. 8 isogeneic hermaphrodite worms were analyzed by serial-section electron microscopy. (A) Schematic of feedforward, feedback and recurrent connections defined by cell types. (B) Proportions of the total number of synapses in feedforward, feedback and recurrent connections. *P = 0.017, ***P = 2.0 × 10^{-4}, NS, not significant (P = 0.11), Spearman's rank correlation, FDR-adjusted using Benjamini–Hochberg correction. (C) Left, schematic of brain-wide synaptic changes from birth to adulthood. Right, principles of maturation describing synaptic changes at the level of brain geometry, individual neurons, neuron types and entire networks. Thicker lines represent stronger connections with more synapses. About 43% of all cell–cell connections—accounting for 16% of all chemical synapses—are not conserved between isogenic individuals.

Credit: Fig. 3a,b and Fig. 4 of ref. 11.

Connections increase as the physical contacts grow, adding synapses to existing connections between neurons and creating new connections between neurons; A portion of the synapses have been remodeled to form subnetworks, and the modules are reminiscent of those in the brain of larger animals (Fig. 3.5C, Chapter 1).

In guinea pig, the number of synapses reaches adult level at birth, but some of the postsynaptic dendritic spines grow much sturdier as the animal matures (Fig. 3.6). We'll see more of these dendritic spines as we discuss memory in Chapter 5.

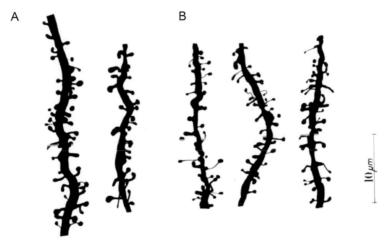

Figure 3.6 Fragments of spiny dendrites of adult guinea pigs (A) and guinea pigs around birth (B). Camera lucida tracings of Golgi pictures. The thicker spines are more frequent in adult samples. So although the number of cortical synapses increased to adult level before birth, we can safely assume that the newborns are actively learning.

Credit: From Fig. 55 of ref. 2, which was Fig. 4 of ref. 12.

In humans, the brain's gray matter volume peaks at 5.9 years while white matter volume peaks at 28.7 years (Fig. 3.7), with potential relevance for age-dependent occurrence of a number of neuropsychiatric disorders[13].

People are better at navigating in environments similar to where they grow up, e.g., cities with a regular layout versus cities with a higher entropy in their street networks[14] (more on spatial navigation in Chapter 7).

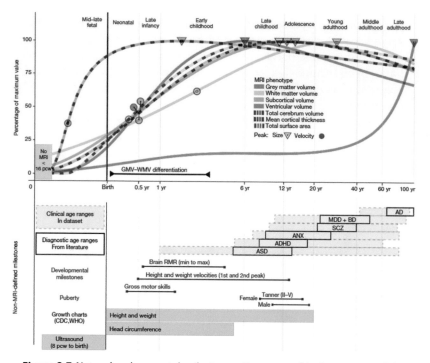

Figure 3.7 Neurodevelopmental milestones. Top, a graphical summary of the normative trajectories of the median (50th centile) for each global magnetic resonance imaging (MRI) phenotype, and key developmental milestones, as a function of age (log-scaled). Circles depict the peak rate of growth milestones for each phenotype (defined by the maxima of the first derivatives of the median trajectories). Triangles depict the peak volume of each phenotype (defined by the maxima of the median trajectories)[13]. Bottom, a graphical summary of additional MRI and non-MRI developmental stages and milestones. From top to bottom: blue shaded boxes denote the age range of incidence for each of the major clinical disorders represented in the MRI dataset; black boxes denote the age at which these conditions are generally diagnosed as derived from literature[13]; brown lines represent the normative intervals for developmental milestones derived from non-MRI data, based on previous literature and averaged across males and females (Methods); gray bars depict age ranges for existing (World Health Organization (WHO) and Centers for Disease Control and Prevention (CDC)) growth charts of anthropometric and ultrasonographic variable. Across both panels, light gray vertical lines delimit lifespan epochs (labelled above the top panel) previously defined by neurobiological criteria. Tanner refers to the Tanner scale of physical development. AD, Alzheimer's disease; ADHD, attention deficit hyperactivity disorder; ASD, autism spectrum disorder (including high-risk individuals with confirmed diagnosis at a later age); ANX, anxiety or phobic disorders; BD, bipolar disorder; MDD, major depressive disorder; RMR, resting metabolic rate; SCZ, schizophrenia.

Credit: Fig. 3 of ref. 13.

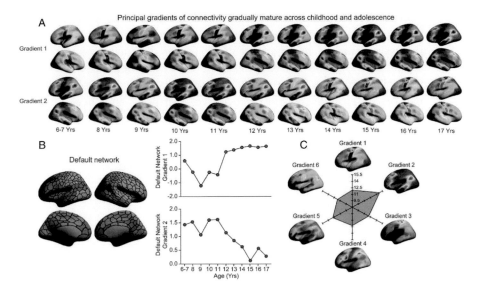

Figure 3.8 Functional connectivity gradients gradually mature across childhood and adolescence (compare with Chapter 1). (A) Participants are divided into groups at 1-year age intervals. Gradient values are displayed on the lateral surface of the left and right hemispheres in which the proximity of colors within each age group indicates the similarity of connectivity patterns across the cortex. The principal gradient (gradient one), which accounts for the greatest variance in connectivity, segregates unimodal regions in childhood. Across development, a gradual transition was observed in gradient one, reflecting the shift to an adult-like architecture after 12 years of age, at one end anchored by sensory and motor regions and at the other end by the association cortex. Gradient two surface maps reveal the inverse transitional profile across development. Here, the extreme ends of gradient two reflect a mixture of the unimodal and association cortex until 14 years old, at which point gradient values within the association cortex decrease, while the functional separation between the somato/motor and visual regions becomes more prominent. (B) Red parcels denote regions within the default network. The black lines denote parcel and network boundaries based on Yeo *et al*.'s[17] seven-network solution averaged across the 400-parcel functional atlas of Schaefer and colleagues[18]. Graphs display average gradient one and gradient two values within the default network for each 1-year bin. (C) Radar plot displays the age at which the default network gradient values peak during development across each of the first six gradients. Radar plots for the remaining large-scale networks are available in ref. 19.

Credit: Fig. 3 of ref. 19.

The human brain's default network (seen as increased blood flow in functional magnetic resonance imaging (fMRI)), is deactivated by attention-demanding tasks that engages the external world, and becomes active for self-referential processes such as autobiographical memory, introspection, emotional processing, and social inferences[15, 16] (more on social cognition in Section 7.9). The external-oriented and the internal-oriented networks appear to compete[15]. The default network probably matures later than visual associative learning, and becomes similar to that in adults only after puberty (Fig. 3.8).

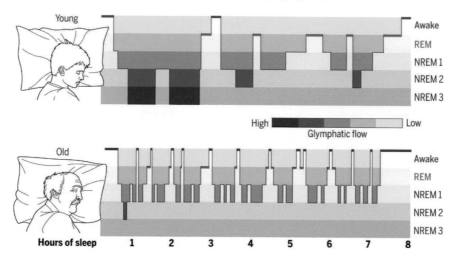

Figure 3.9 Sleep architecture in young and old individuals. Hypnograms are constructed from EEG recordings and display the cyclic transitions between sleep stages. The two schematic hypnograms illustrate the sleep architecture of young and old individuals who transition spontaneously between the awake state, REM sleep, and NREM (stages 1 to 3) sleep. Stage 1 NREM sleep is light sleep, whereas stage 3 NREM sleep is the deepest sleep stage and is characterized by slow-wave EEG activity. For young people, deep (stage 3) NREM sleep dominates in the early phases of sleep, whereas REM sleep is more frequent in the later phases. Sleep spindles are most frequent in stage 2 NREM sleep. By contrast, for people older than 60 years of age, sleep is often interrupted by short awake episodes, and older individuals do not typically enter stage 3 NREM sleep. Total sleep time decreases by 10 min for each decade of life[20]. Green shading indicates the proposed efficacy of glymphatic clearance on the basis of data collected in rodents[24, 25]. The lack of stage 3 NREM sleep, the frequent interruptions of stage 1 and 2 NREM sleep, and the shorter total sleep time all serve to decrease glymphatic activity in aging. Critically, a number of disorders and conditions can suppress glymphatic function during NREM sleep, further exacerbating the effects of glymphatic dysfunction in neurodegenerative disease.

Credit: Fig. 3 of ref. 26.

This default network has been functionally linked to the anterior hippocampus, while the posterior hippocampus is linked to the parietal memory network (PMN) including visual/perceptual neocortical regions[16], similar to the ventral versus dorsal hippocampus in rats and mice (Figs. 1.6 and 6.12).

With all the neuronal connections essentially there, older people do not sleep as much or as deeply (Fig. 3.9), but could function more robustly despite sleep deprivation[20,21]. In aged mice, hypocretin/orexin-expressing neurons associated with wakefulness and REM (Rapid Eye Movement) sleep were fewer in number, but fired more frequently[22, 23]. Older deer and elephants can lead their group to sources of food or water in time of crises. So the memory (Chapters 4 and 5) is perhaps not overwritten for a reason. We'll talk more about sleep in Chapter 6.

3.3   Summary

As a book for electrical engineers and data scientists first, it would be too much to talk about various aspects of neurodevelopment. Young animals are perhaps where the traditional term of "wiring" is appropriate (but not in worms or guinea pigs, who already have their blueprint laid out), with cells being divided up, moving into places and forming landmarks, branching out and figuring out their connections. This is perhaps one reason why psychiatrists dig into one's childhood experiences. As Dr. Santiago Ramón y Cajal said, "the total arborization of a neuron represents the graphic history of conflicts suffered from its embryonic life"[27]. From an engineering point-of-view, things always have to work under limited resources. Some people may devote more cells, and branch more readily to specific tasks. How we age without becoming an overfit model is likely a global challenge.

Questions

1. As scientists profile each type of cells in the brain of a model animal, how much variation do we expect to see among individuals of the same species?

2. Together with information from the previous chapters, what do you think would be good initial parameters for neurons in a newborn? Would the answers be different in a bad year?

References

1. CE, C. & RE, B. Development of the time coding pathways in the auditory brainstem of the barn owl. *J. Comp. Neurol.* **373**, 467–483 (1996).

2. Braitenberg, V. & Schüz, A. Cortex: statistics and geometry of neuronal connectivity. *Cortex Stat. Geom. Neuronal Connect.* (1998) doi:10.1007/978-3-662-03733-1.

3. Sathyanesan, A. *et al.* Emerging connections between cerebellar development, behaviour and complex brain disorders. *Nat. Rev. Neurosci.* **20**, 298–313 (2019).

4. Branco, T. & Staras, K. The probability of neurotransmitter release: variability and feedback control at single synapses. *Nat. Rev. Neurosci.* **10**, 373–383 (2009).

5. Paulsen, B. *et al.* Autism genes converge on asynchronous development of shared neuron classes. *Nature* **602**, 268–273 (2022).

6. Paredes, M. F. *et al.* Nests of dividing neuroblasts sustain interneuron production for the developing human brain. *Science (80-.).* **375**, eabk2346 (2022).

7. Delgado, R. N. *et al.* Individual human cortical progenitors can produce excitatory and inhibitory neurons. *Nature* **601**, 397–403 (2022).

8. Zhang, T. *et al.* Generation of excitatory and inhibitory neurons from common progenitors via Notch signaling in the cerebellum. *Cell Rep.* **35**, 109208 (2021).

9. Wong, F. K. *et al.* Pyramidal cell regulation of interneuron survival sculpts cortical networks. *Nature* **557**, 668–673 (2018).

10. Biran, V., Verney, C. & Ferriero, D. M. Perinatal cerebellar injury in human and animal models. *Neurol. Res. Int.* **2012**, 858929 (2012).

11. Witvliet, D. *et al.* Connectomes across development reveal principles of brain maturation. *Nature* **596**, 257–261 (2021).

12. Schüz, A. Prenatal development and postnatal changes in the guinea pig cortex: microscopic evaluation of a natural deprivation experiment. II. Postnatal changes. *J. Hirnforsch.* **22**, 113–27 (1981).

13. Bethlehem, R. A. I. *et al.* Brain charts for the human lifespan. *Nature* **604**, 525–533 (2022).

14. Coutrot, A. *et al.* Entropy of city street networks linked to future spatial navigation ability. *Nature* (2022) doi:10.1038/s41586-022-04486-7.

15. Buckner, R. L. & DiNicola, L. M. The brain's default network: updated anatomy, physiology and evolving insights. *Nat. Rev. Neurosci.* **20**, 593–608 (2019).

16. Zheng, A. *et al.* Parallel hippocampal-parietal circuits for self- and goal-oriented processing. *Proc. Natl. Acad. Sci. U. S. A.* **118**, e2101743118 (2021).

17. Yeo, B. T. T. *et al.* The organization of the human cerebral cortex estimated by intrinsic functional connectivity. *J. Neurophysiol.* **106**, 1125–65 (2011).

18. Schaefer, A. *et al.* Local-global parcellation of the human cerebral cortex from intrinsic functional connectivity MRI. *Cereb. Cortex* **28**, 3095–3114 (2018).

19. Dong, H.-M., Margulies, D. S., Zuo, X.-N. & Holmes, A. J. Shifting gradients of macroscale cortical organization mark the transition from childhood to adolescence. *Proc. Natl. Acad. Sci. U. S. A.* **118**, e2024448118 (2021).

20. Landolt, H. P. & Borbély, A. A. Age-dependent changes in sleep EEG topography. *Clin. Neurophysiol.* **112**, 369–77 (2001).

21. Taillard, J., Gronfier, C., Bioulac, S., Philip, P. & Sagaspe, P. Sleep in normal aging, homeostatic and circadian regulation and vulnerability to sleep deprivation. *Brain Sci.* **11**, 1003 (2021).

22. Feng, H. *et al.* Orexin signaling modulates synchronized excitation in the sublaterodorsal tegmental nucleus to stabilize REM sleep. *Nat. Commun.* **11**, 3661 (2020).

23. Li, S.-B. *et al.* Hyperexcitable arousal circuits drive sleep instability during aging. *Science (80-.).* **375**, eabh3021 (2022).

24. Xie, L. *et al.* Sleep drives metabolite clearance from the adult brain. *Science* **342**, 373–7 (2013).

25. Hablitz, L. M. *et al.* Increased glymphatic influx is correlated with high EEG delta power and low heart rate in mice under anesthesia. *Sci. Adv.* **5**, eaav5447 (2019).

26. Nedergaard, M. & Goldman, S. A. Glymphatic failure as a final common pathway to dementia. *Science (80-.).* **370**, 50–56 (2020).

27. Ehrlich, B. *The Brain in Search of Itself : Santiago Ramón y Cajal and the Story of the Neuron.* (Farrar, Straus and Giroux, 2022).

Chapter 4

Memory in Cells

Abstract

We continue to be fascinated by how much and how efficiently animals and humans can remember. The physical basis of memory is now increasingly clear. Before neuroscience researchers agree with each other, the cell-based models for memory and learning presented here and in the next few Chapters already have great potential for the next generation of machine-learning algorithms that would never stop learning. One key question faced by the brain with the daily bombardment of information is what to store and where to store. We begin to understand its ingenious solution to the problem, including how new dendrites or new neurons are added.

Keywords

Memory engram, Volume transmission, Turing pattern, Memory consolidation, Hashing, Abstract learning, Hippocampus, Adult neurogenesis, Forgetting

4.1 Engrams: Single-Cell Basis of Memory

In recent years, it has become clear that there is nothing meta-physical about memory in the brain. So it comes down to the cells and how they are organized (Figs. 4.1 and 4.2). With extensive

Neuroscience for Artificial Intelligence
Huijue Jia
Copyright © 2023 Jenny Stanford Publishing Pte. Ltd.
ISBN 978-981-4968-78-2 (Hardcover), 978-1-003-41098-0 (eBook)
www.jennystanford.com

experience, many neurons linked to different senses could involve one's grandma, so that we don't need to worry about losing a single "Grandma cell" that goes unrenewed.

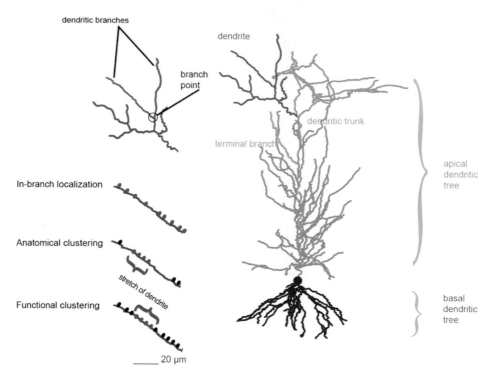

Figure 4.1 Dendritic structure and plasticity. Each dendritic tree (apical or basal) in pyramidal neurons can be subdivided to a number of dendrites (dendritic subtrees connected to the apical trunk or the soma). Thin terminal branches are the main targets of excitation in the cerebral cortex. There, synaptic inputs can be organized in the following ways[1]: (1) they can be localized in the same dendritic branch without specific spatial arrangement (in-branch localization), (2) they can form anatomical clusters, whereby spines form morphologically distinct groups of several spine heads located in distances less than 5 microns from each other within stretches of a given branch and (3) they can form functional clusters where spine density is uniform but nearby synapses (located within 10–20 µm) are activated synchronously (see also refs. 2 and 3). The implications of these different arrangements of connectivity at the dendritic level are discussed in Section 4.2.

Credit: Fig. 1 of ref. 1.

Pyramidal neurons go across layers. The majority of excitatory synapses are from one pyramidal cell to another pyramidal cell[4].

Somewhat like the tentacles of a grape vine, related branches of a dendrite can form synapses with multiple neurons, and multiple neurons can form neighboring synapses on the same neuron (Fig. 4.1). Proteins such as NMDA (N-methyl-D-aspartate) receptors (Fig. 4.3), AMPA (α-amino-3-hydroxy-5-methyl-4-isoxazole propionic acid) receptors, actin filaments[5-8], determine the strength of each synapse (more in Chapter 5). Concentrations of such resources can be both local and global, and epigenetics impact the supply of mRNA and ribosomes from the cell body (soma).

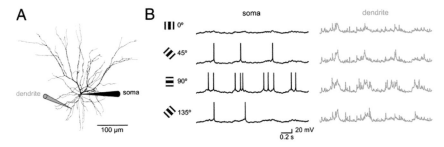

Figure 4.2 A pyramidal cell model with active dendrites reproduces dendritic responses to visual stimulation in vivo. (A) Morphology of the L2/3 pyramidal neuron used for the model, with pipettes indicating the locations of the recordings shown in B. (B) Example of somatic (Left) and dendritic (light blue, Right) responses of the postsynaptic model neuron to simulated synaptic inputs corresponding to visual stimuli with orientations 0°, 45°, 90°, and 135°, where 90° is the preferred orientation of the model neuron.

Credit: Fig. 1A,B of ref. 9.

4.2 To Engage More Cells for a Stronger Memory?

Besides the classic point-to-point neural transmission through synapses in milliseconds time scale, on the faster side, there is dendritic gap junctions (in nanoseconds, good for coordination of interneurons[10]), and on the slower side, there is volume transmission by diffusible molecules or vesicles[11,12]. With protein leakage to nearby synapses and volume transmission, it might be possible to generate Turing patterns in the brain (Figs. 4.4 and 4.5)[13, 14]: circles or stripes formed by local activation, and inhibition from afar in a Reaction-Diffusion system. Such waves of diffusion or electric signals would also be one way of having a temporal order (Section 7.3). Medication such as monoamine oxidase inhibitors (MAOi) used to treat depression might allow the circles to get larger (Fig. 4.5), so that new activities could become more memorable.

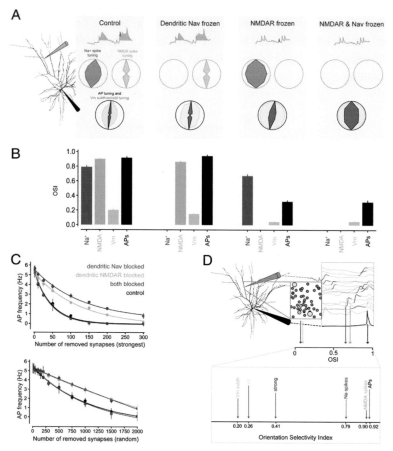

Figure 4.3 Dendritic spikes enhance control of action potential (AP) output and its tuning by a small number of strong synaptic inputs. (A) Example of the tuning of dendritic Na+ spikes (red) and NMDA spikes (orange) as well as subthreshold somatic membrane potential (gray) and APs (black), under control conditions and with dendritic Na+ spikes (second from Left), NMDA spikes (third from Left), or both (Right) blocked. (B) Mean Orientation Selectivity Index (OSI) of dendritic Na+ spikes (red), NMDA spikes (orange), somatic Vm (gray), and APs (black) under control conditions (Left), with Nav channels blocked (second from Left), with NMDA conductances frozen (third from Left), or both (Right). (C, Top) Removing strong synapses decreases AP

(*Continued*)

Figure 4.3 (*Continued*)

frequency in response to a preferred visual stimulus in the full model (black) and under Nav channel block (red). The decrease can be fit with an exponential decay and is less steep when NMDARs only (orange) or NMDARs and Nav channels (purple) are blocked. Constant current was injected at the soma to maintain AP frequency at control levels with zero synapses removed. This was done to engage the somatic output nonlinearity at the same starting point as in control. (C, Bottom) Removing random synapses decreases AP frequency in response to a preferred visual stimulus in the full model (black) and under Nav channel block (red). The decrease can be fit with an exponential decay. When NMDARs only (orange) or NMDARs and Nav channels (purple) are blocked, the decrease can be fit linearly and is less steep. (D, Top) Schematic of the composition of spatiotemporal STAs of synaptic activity and a spatiotemporal AP-triggered average of dendritic spikes. These were used to dissect the spatiotemporal constellations of synaptic inputs that evoke dendritic spikes and the effect of such dendritic spikes on AP output. (D, Bottom) OSI increases from synaptic input via dendritic spikes to AP output. The OSI of synaptic inputs (light blue: all synapses; dark blue: strong synapses), dendritic Na$^+$ spikes (red), and NMDA spikes (orange) as well as subthreshold somatic membrane potential (gray) and APs (black) are shown.

Credit: Fig. 6 of ref. 9.

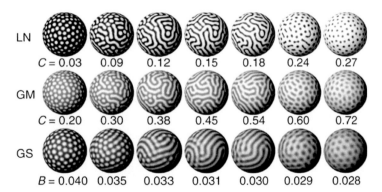

Figure 4.4 Two-dimensional Turing patterns generated by numerical simulations. In each variation of the Reaction-Diffusion model originally proposed by Alan Turing[13], gradual modulation of one parameter value (indicated below each sphere) can cause pattern changes from spots (left most) to inverse spots (right most). Intermediate regions always appear as labyrinthine stripe patterns. Each color represents the concentration of the core factor in each model (brown: activator in the linear model, LN; orange: activator in the Gierer–Meinhardt model, GM; blue: autocatalytic enzyme in the Gray–Scott model, GS. Lighter color indicates higher concentration). The equations and details for simulations are provided in the Methods section of ref. 15.

Credit: Fig. 1 of ref. 15.

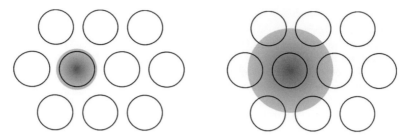

Figure 4.5 Range or waves of a signal might determine the number of neurons that are allocated to each event. For example, a very emotional event could recruit too many neurons. Medication that changes the local signal or the inhibition could modulate the patterns.

Credit: Huijue Jia.

4.3 Competition for Allocation into a Memory Engram

New experiences are either written onto neurons that have previously encoded a related experience, or written to new places (Sections 4.4 and 4.5; more on the recruitment in Chapter 9, more on sleeping in Chapter 6). Whichever neurons that happened to be more active at the time could participate in the memory. Such a competitive encoding process naturally leads to fewer cells and synapses that represent distant events, while some events occupied many more cells than other events to begin with (Fig. 4.5), and also tend to replay (Chapter 5).

Not only events, but also rules are memorized in the prefrontal cortex and updated at the single-cell level (Figs. 4.2, 4.6, 4.7 and 4.8), and likely in individual dendritic spines (Chapter 5). Even after all the connections are lost, as long as the neurons and the proteins are still there, the memory engram could be directly activated (a shine of laser on permanently labelled neurons in mice models)[16, 17]. In that sense, we may indeed remember everything that we ever remember; it is more a problem of retrieval.

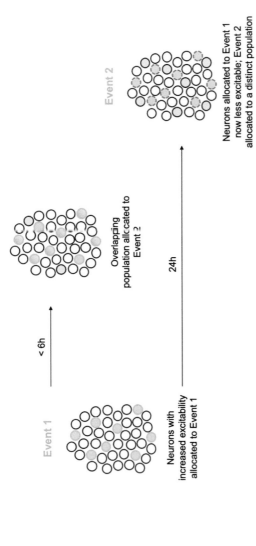

Event 1

< 6h

Neurons with
increased excitability
allocated to Event 1

Event 2

Overlapping
population allocated to
Event 2

24h

Event 2

Neurons allocated to Event 1
now less excitable; Event 2
allocated to a distinct population

Figure 4.6 Neuronal allocation and memory linking[16]. Neurons with increased excitability at the time of event 1 (blue) are allocated to the engram supporting this memory (blue filled circles outlined in orange). These allocated engram neurons remain more excitable than their neighbors for several hours after event 1. If a similar event 2 (green) occurs during this time, neurons allocated to the engram supporting event 1 are more excitable and, therefore, also allocated to the engram supporting event 2 (blue and green filled circles outlined in orange). In this way, neurons are co-allocated to events 1 and 2. By virtue of co-allocation, these two memories become linked. After some time, neurons allocated to the engram supporting event 1 become less excitable than their neighbors ("refractory"), and if event 2 occurs in this time window, a new population of more excitable neurons wins the competition for allocation to the engram supporting event 2. This disallocation allows the two memories to be remembered separately. Circles with red dashed outlines represent less excitable neurons.

Credit: Fig. 5 of ref. 16.

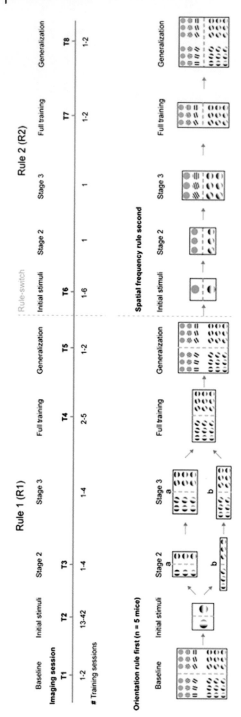

Figure 4.7 Timeline of behavioral training and presented stimuli of individual mice for categorization rules. The 36 gratings used as stimuli for the mice differed in orientation and spatial frequency. Up to 18 gratings were introduced in a stepwise training, according to either the orientation rule first or the spatial frequency rule first. Switch to the second rule was also trained in steps. The mice were randomly assigned to get trained on either of the rules first. (Upper panel) Timeline showing behavioral training stages, the number of training sessions that mice spent in each stage (min–max) and the imaging sessions (T1–T8). (Lower panel) Stimuli used for category training, aligned to the stages shown in in the upper panel. The scheme shows stimuli for mice that were trained on the orientation rule first, and the spatial frequency rule second.

Credit: Part A and B, Extended Fig. 1a and 1c, ref. 18.

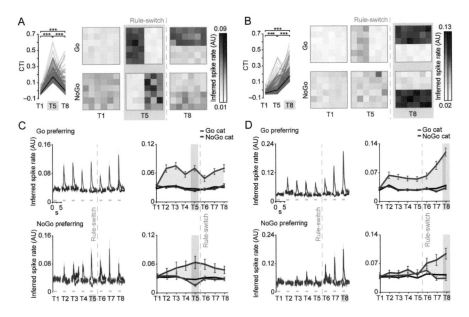

Figure 4.8 Two populations of category-selective neurons in the prefrontal cortex show different dynamics during a rule switch. (A) Left, CTI (Category-tuning index, with values from 0 to 1 indicating no difference to strong differences in activity between, but not within, categories) of all category-selective neurons identified at T5 (gray highlight), shown for time points T1, T5 (first rule) and T8 (second rule, Fig. 4.9). $P_{T1-T5} = 1.1 \times 10^{-36}$, $P_{T1-T8} = 1.6 \times 10^{-14}$, $P_{T5-T8} = 6.9 \times 10^{-27}$, two-tailed WMPSR test, Bonferroni-corrected for three comparisons (n = 213 cells). Black line denotes the mean. Right, average inferred spike rate per stimulus of Go and NoGo category-selective cells at T1, T5 and T8 (neurons that activated when the mice chose to lick for the water reward or not: nGo = 156 cells; nNoGo = 57 cells) (B) As in (A), but for category-selective cells defined at T8 (after the rule switch, Fig. 4.9). $P_{T1-T5} = 4.2 \times 10^{-18}$, $P_{T1-T8} = 2.9 \times 10^{-33}$, $P_{T5-T8} = 1.1 \times 10^{-27}$, two-tailed WMPSR test, Bonferroni-corrected for three comparisons (n = 192 cells; nGo = 122 cells; nNoGo = 70 cells). (C) Left, inferred spike rate of all Go (top) and NoGo (bottom) category-selective cells, identified at T5 (gray highlight), during trials of all Go (green) and NoGo (red) category stimuli at T1–T8. Gray denotes stimulus presentation. Data are mean ± s.e.m., across cells. Right, inferred spike rate during stimulus presentation of all Go (top) and NoGo (bottom) category-selective cells. Green denotes Go category, red denotes NoGo category, orange area denotes the difference. Black denotes the mean inferred spike rate in the pre-stimulus period. Data are mean ± s.e.m., across cells. (D) As in (C), for category-selective cells defined at T8.

Credit: Fig. 3 of ref. 18.

4.4 Memory Consolidation in View of Hashing

The classic view for memory consolidation is that the hippocampus (Figs. 4.9, 1.5 and 1.6) is a temporary storage that gradually transmits "information" for permanent storage in the cortex (Fig. 4.10). However, like what computers already do, but with feedback from the cortex after a search (hippocampal θ waves modulate cortical γ waves[19,20]; hippocampal sharp-wave ripples preceded cortical ripples during NREM sleep[21] (Chapter 6)), the hippocampal pyramidal cells (place cells) might constitute a Hashing function that writes into the cortex at the very moment (Fig. 4.11). The slower waves could represent a broader search with groups of oscillating neurons including interneurons (Section 6.3), while specific memory is retrieved by sharp-wave ripples[22-27] (SWRs, Chapter 5). Hippocampal activity, for example, was responsible for coherent boundaries between events, which was reactivated when telling a remembered story the next day[28, 29].

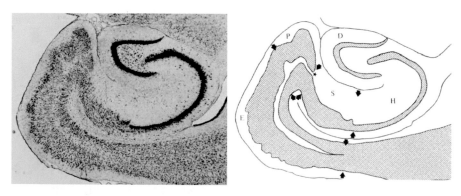

Figure 4.9 Continuation of the cortex into the mouse hippocampus. Left, Nissl preparation; right, schematic chart. It can be seen that of the two main layers of the entorhinal cortex (E), the more superficial one ends (asterisk) in the region of presubiculum, while the lower one turns into the subiculum (S), whose cell layer then flattens to form the dense band of pyramidal cells in the hippocampus (H). The arrows indicate the borders between these regions. The dentate gyrus (D) is a separate sheet. This is Nissl staining of the mouse cortex, please note the different position of the MEC in monkeys and in humans (Fig. 1.6).

Credit: Fig. 75 of ref. 4, from ref. 30.

Cortical modules

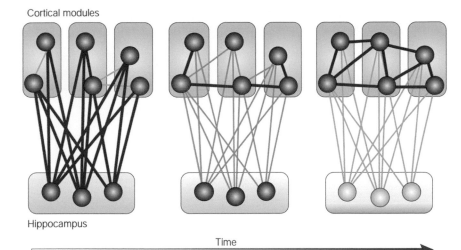

Hippocampus

Time

Figure 4.10 Standard consolidation model. Encoding of perceptual, motor and cognitive information initially occurs in several specialized primary and associative cortical areas. The hippocampus integrates information from these distributed cortical modules that represents the various features of an experience, and rapidly fuses these features into a coherent memory trace[31, 32]. Successive reactivation of this hippocampal–cortical network leads to progressive strengthening of cortico-cortical connections (for example, by strengthening existing cortico-cortical connections or establishing new ones). Incremental strengthening of cortico-cortical connections eventually allows new memories to become independent of the hippocampus and to be gradually integrated with pre-existing cortical memories[33, 34]. A key feature of this model is that changes in the strength of the connections between the hippocampal system and the different cortical areas are rapid and transient, whereas changes in the connections between the cortical areas are slow and long-lasting[33, 34].

Credit: Fig. 1 of ref. 35.

The weeks of delay for memory consolidation could be explained by incorporation of new neurons (Section 4.5) and stabilization of new synapses, including during sleep. Over time, the hippocampal representation decreased in temporal resolution (Fig. 5.14; Section 7.3), and both the hippocampal cells and the cortical cells could have accumulated links to other experiences (Fig. 4.6). Not just the same cells, but the same region of a dendritic branch would be used for the same task, e.g., olfactory memory in fruit fly[36], pyramidal cells in the motor cortex[37, 38], consistent with local computation of the dendritic spines (Chapter 5).

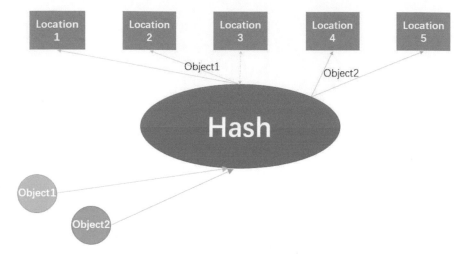

Figure 4.11 Feedback hashing in the brain. After a search (e.g., driven by the claustrum-MEC-hippocampus, or by the amygdala-ACC-hippocampus), existing representation of a place or an event is generated with stored information from the cortex (Chapter 7, and Fig. 9.7). The input from the cortex to the hippocampus is computed within the dendrites (Chapter 5), which may or may not lead to firing of the hippocampal pyramidal cell (Fig. 4.2). New synapses are sculpted as this process plays out a few times (Chapter 6 on sleep), including recruitment of new branches and new cells (dashed arrow for object 1). The hash function is thereby improved with new experience, consistent with Hebbian learning.

Credit: Huijue Jia.

I suppose that the cortical place for memory storage (more in Chapter 9) can be uniquely defined by one keyword (e.g., people, place) plus temporal order, or by two or more keywords together. Interestingly, this likely hippocampus-mediated searching and hashing function only appeared mature after adolescence[39] (Section 3.2). We'll fill in key parts of such a feedback hashing algorithm (Figs. 4.11 and 4.12) after discussing replays (Chapter 5), sleep (Figs. 6.6 and 6.11) and then space and time (Chapter 7).

Contrary to common belief that the hippocampus would no longer be involved in memory retrieval after it has "transmitted information" to the cortex (Fig. 4.10), normal activity of hippocampal CA1 (cornu ammonis region 1) neurons was required for fast retrieval of remote memory[40]. With very high activity in the anterior cingulate cortex (ACC, often studied for chronic pain but likely more about decisions and rewards through the amygdala[41-43]), however, retrieval of the remote memory is possible with pharmacologically inhibited (30 min) hippocampal CA1 neurons[40].

Figure 4.12 Emergence of pattern after a small number of experiences. More on higher level cognition in Chapters 8 and 9.

Credit: Huijue Jia.

Besides direct connections between the ACC and hippocampus (Section 5.3), the subiculum (Fig. 4.9), CA1 and CA3 of the hippocampal formation are linked to the anterior thalamic nuclei of the thalamus, which has reciprocal projections with the ACC[44]. The anterior thalamic nuclei also contain head direction cells[44] (Chapter 7), important for spatial navigation.

4.5   Combining Old and New

Addition of new neurons is more than simple addition of computational nodes and searching indices. In food-storing birds, the size of the hippocampus increases every autumn and winter, due to neurogenesis[45]. In rats, new neurons are gradually added from the subventricular zone (SVZ, which generates olfactory neurons in adults, but was an important source of neurons early in development[46]) and the dentate gyrus of hippocampus throughout life[47, 48] (Figs. 4.13 and 5.14), but the number decreases with age[49]. The monounsaturated fatty acid, oleic acid, for example, has been found to promote hippocampal neurogenesis[50].

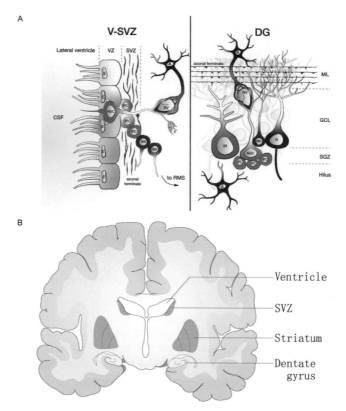

Figure 4.13 The adult neural stem cell niche from which new neurons emerge. (A) Schematic representation of the niches in the rodent (typically rats or mice) ventricular-subventricular zone (V-SVZ) and dentate gyrus (DG). Neural stem cells in these niches are active throughout life and continuously produce new neurons that integrate into the pre-existing neuronal network as well as astrocytes and oligodendrocytes. Adult neural stem cells in these two locations share many characteristics but are also regulated by different signals coming from their distinct environments. For instance, V-SVZ NSCs are directly regulated by factors present in the cerebrospinal fluid, while DG NSCs receive feedback from the neurons they generate, which settle in the granule cell layer. A, astrocytes; BV, blood vessel; CSF, cerebrospinal fluid; E, endothelial cell; F, fractone; GCL, granule cell layer; IN, interneuron; IPC, intermediate progenitor cell; ML, molecular layer; N, neuron; NB, neuroblast; NSC, neural stem cell; RMS, rostral migratory stream; SGZ, subgranular zone; SVZ, subventricular zone; VZ, ventricular zone. (B) A coronal view of the human brain highlighting areas where neurogenesis has been described in the human brain. The DG might not be the major site of hippocampal neurogenesis in humans[54]. For details regarding the controversy and evidence for each site, see refs. 48 and 62.

Credit: Part A, from Fig. 3 of ref. 47; Part B, the right panel of Fig. 1[48].

Adult neurogenesis in humans is more difficult to study and has remained controversial[48, 51-53]. Expression of doublecortin (*DCX*), a widely used marker for neuroblasts (progenitor cells for the neurons) and new dentate gyrus granule cells in rats and mice, was found in many cells in the hippocampal CA1-CA4 in pigs, monkeys and humans, including interneurons[54] which might involve new types that do not exist in mice[55] (more on interneurons in Sections 6.3, 6.5, and 7.8). Microglia, at least during development, regulate entry of parvalbumin-expressing interneurons into circuits[56]. Adult neurogenesis decreased in patients with mesial temporal lobe epilepsy, while production of astrocytes persisted[57].

Apolipoprotein E (ApoE) release from astrocytes appeared required for proper maturation of new hippocampal neurons to obtain a normal density of dendritic spines[58, 59], which is likely a universal problem for Alzheimer's disease. The dendritic spines are larger with the ApoE version that predispose people to Alzheimer's disease (e.g., due to higher 27-hydroxycholesterol[60]); and larger spines are presumably more persistent (Chapter 5). Moreover, the converge point of dendritic branches looked closer to the neuronal cell body[58, 59], so the cell body might more easily become active.

During aging (and perhaps other inflammatory conditions), immune cells infiltrate the SVZ and dentate gyrus. Natural killer cells, in particular, eliminate aged neuroblast cells in the dentate gyrus, thereby contributing to the age-related decline in neurogenesis and cognitive function[61]. Acute mobilization of the stem cell pool, such as after stroke or severe stress, can mean long-term depletion[47].

According to radiocarbon dating, about 700 new neurons are born into the hippocampus each day in a middle-aged human, which add up to a 1.75% annual turnover of the 1/3 hippocampal neurons that renew, whereas the nonrenewing neurons die without being replaced[51]. The half-life of a neuron in the renewing fraction is 7.1 years—10× shorter than in the nonrenewing fraction[51].

The new neurons mature and get integrated into neuronal circuits in the following weeks, being easier to activate than mature neurons[48, 63, 64]. With the 2–3-week delay in rats between the emergence of new neurons and their integration into circuits (Fig. 5.14), I guess the new neurons would need to be tagged upon birth (e.g., epigenetic choice of adhesion molecules), together with branches of the existing circuit that are insufficient for the new experience.

New dendritic spines grow towards sources of neuro-transmitters such as glutamate[65]. Studies in mice or rats showed two waves of partial elimination of newborn neurons: the first wave within 48 h of neurogenesis, the second wave 12–16 days after neurogenesis, when new neurons fail to receive NMDA receptor-dependent input and integrate into existing circuits[48].

Neurogenesis facilitates the formation of new memory[66, 67], whereas existing memory can be forgotten[68] (not retrievable with everyday cues, Fig. 4.11). New neurons added to the posterior (dorsal in rats) hippocampus can facilitate cognition and spatial learning; new neurons added to the anterior (ventral in rats) hippocampus can mediate emotional behavior, social interactions, and stress resilience[69] (reminiscent of the default network in Section 3.2). Interestingly, the anterior hippocampal place cells have larger place fields (e.g., 10 m for a rat, Fig. 6.12) than the posterior place cells, i.e., lower resolution (more on spatial cognition in Chapter 7).

For conscious memory replays, I would favor the idea that the hippocampus keeps a sparse set of the threads with the cortex even after memory consolidation (Figs. 4.10 and 4.11), while the sparsity prevents complete overwriting of old information in the cortex. It follows that a straight-forward way of encoding new experience is to add a few new neurons. As we may know from teaching students, to refer to some familiar examples would likely facilitate proper hashing to existing neural circuits (Fig. 4.12, Chapter 9), potentially updating the old memory and saving some new neurons for other activities.

In a study of rats exposed to a new maze (Fig. 4.14), it was shown during the sleep after the maze exploration that many of the hippocampal place cells that fired during the sleep before the new maze became activated in a new sequence when asleep, in the same new order as during the exploration while awake (Fig. 4.15). We may guess that the rats recall previous experiences, and only add new neurons to fill in the differences for the new maze (Fig. 5.14); The same can be said for categorical learning (Fig. 4.8). In a novel environment, the setting on hippocampal place cells in the CA1 layer is generally loosened by interneurons[70], to more easily get a match with existing cells; and the senses are likely more keen (e.g., due to feedback from the claustrum, Fig. 7.15B), to effectively add new nodes where necessary.

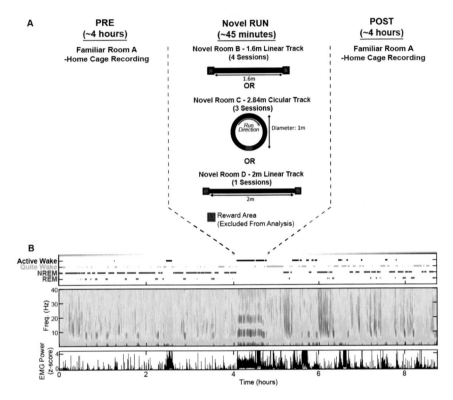

Figure 4.14 Behavioral and recording paradigm. All sleep and novelty recordings were performed during the animal's day-cycle when rats are prone to sleep. (A) Each recording session consisted of three phases: (1) a long (~4 h) PRE epoch in which the animal was allowed to rest or sleep in its homecage in familiar room A, (2) the animal was subsequently transferred to one of three rooms it had never previously been exposed to, and rewarded for running on a novel maze (MAZE), (3) the animal was returned to its homecage in room A for another long (~4 h) POST sleep/rest recording. Note that the reward areas (red) of each maze were excluded from the analyses (the dimensions shown for each maze reflect the linearized lengths after the exclusion of the reward areas). (B) PRE, MAZE and POST epoch sequence content was only assessed during ripple events occurring during quiet immobile waking (green) or NREM sleep epochs (blue) as determined by hippocampal CA1 pyramidal LFP spectrograms and simultaneous electromyographic (EMG) recordings.

Credit: Fig. S1 of ref. 76.

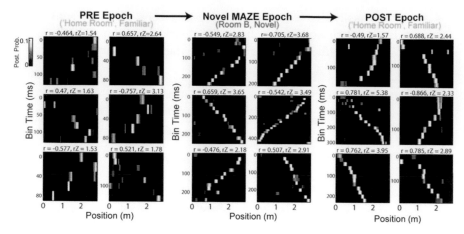

Figure 4.15 Place cell firing sequence events PRE and POST maze (illustrated in Fig. 4.14). Four male rats were implanted with silicon probes for simultaneous recording of place cells in the CA1 layer of the hippocampus. Representative forward and reverse sequences during PRE maze sleep, immobility in the novel MAZE, and POST-learning sleep. A spatial Bayesian decoder[77], constructed from the firing-rate vectors of place cells (n = 491 cells) during track running (eight novel exploration sessions), was applied to all candidate ripple events of PRE, MAZE, and POST immobility epochs to estimate the posterior probabilities of position in forward or reverse virtual traversals of the track. These virtual traversals were measured as weighted correlations over the Bayesian derived posteriors for place across all 20 ms bins in each ripple event and normalized as Z scores.

Credit: Fig. 1B of ref. 76.

New branches can be formed after a few rounds of practice in the dendrites of existing cells[71]. The neck of each dendritic spine, and thereby the synaptic weight, is likely sculpted during such repeated tests from the stubby immature form[5, 72] (Chapter 5). If the stubby spines are not too much off compared to the mature ones, their lack of a neck could mean larger weights for the newcomers in dendritic computation that more easily engage nearby regions; yet we'll see that their own synaptic weights are not increased as effectively (Chapter 5).

Myelination around the axons can be modulated to improve memory[71, 73], including myelination on interneurons and callosal projection neurons (connect between hemispheres through the corpus callosum) in layer 2/3[74], and the white matter could differ wildly in efficiency (Chapter 1). Bidirectional communication between the neocortex and the thalamus can involve long loops[75]. We'll talk more about sleeping and dreaming in Chapter 6, and more on space and time in Chapter 7.

4.6 Summary

In rats, the addition of new neurons, i.e., adult neurogenesis, is best known to happen in the hippocampal dentate gyrus, while it might involve different sites and cell types in monkeys and humans. The ubiquitously cited classic model of memory consolidation states that information is temporarily stored in the hippocampus and is transferred to the neocortex over weeks. Rather, emerging evidence leads me to a more hardware-based model, whereby entorhinal-hippocampal waves and ripples direct a search of existing information in the cortex, and wherever inadequate, encourages storage of new information, with hippocampal pyramidal neurons serving as hash functions. I'll further develop this feedback hashing model after Chapters 5, 6, and 7 (memory in dendritic spines, sleep, and then space and time), and infer its cognitive functions in Chapter 9.

Questions

1. What are the keywords that one typically uses to retrieve memories? How does a slow search compare with a fast search (so fast that one may not even realize)?

2. How do you think we usually allocate the hundreds of new hippocampal neurons every day? How many of them survive into mature forms? How many hippocampal and cortical neurons could be remaining for a distant event?

References

1. G, K., DJ, C., SC, M., AJ, S. & P, P. Synaptic clustering within dendrites: an emerging theory of memory formation. *Prog. Neurobiol.* **126**, 19–35 (2015).

2. Adoff, M. D. *et al.* The functional organization of excitatory synaptic input to place cells. *Nat. Commun.* **12**, 3558 (2021).

3. Megías, M., Emri, Z., Freund, T. F. & Gulyás, A. I. Total number and distribution of inhibitory and excitatory synapses on hippocampal CA1 pyramidal cells. *Neuroscience* **102**, 527–540 (2001).

4. Braitenberg, V. & Schüz, A. Cortex: statistics and geometry of neuronal connectivity. *Cortex Stat. Geom. Neuronal Connect.* (1998) doi:10.1007/978-3-662-03733-1.

5. Crick, F. Do dendritic spines twitch? *Trends Neurosci.* **5**, 44–46 (1982).

6. Okabe, S. Regulation of actin dynamics in dendritic spines: nanostructure, molecular mobility, and signaling mechanisms. *Mol. Cell. Neurosci.* **109**, 103564 (2020).

7. Khanal, P. & Hotulainen, P. Dendritic spine initiation in brain development, learning and diseases and impact of BAR-domain proteins. *Cells* **10**, 2392 (2021).

8. Goto, A. *et al.* Stepwise synaptic plasticity events drive the early phase of memory consolidation. *Science (80-.).* **374**, 857–863 (2021).

9. Goetz, L., Roth, A. & Häusser, M. Active dendrites enable strong but sparse inputs to determine orientation selectivity. *Proc. Natl. Acad. Sci. U.S.A.* **118**, e2017339118 (2021).

10. Szoboszlay, M. *et al.* Functional properties of dendritic gap junctions in cerebellar Golgi cells. *Neuron* **90**, 1043–1056 (2016).

11. Borroto-Escuela, D. O. *et al.* The role of transmitter diffusion and flow versus extracellular vesicles in volume transmission in the brain neural-glial networks. *Philos. Trans. R. Soc. Lond. B. Biol. Sci.* **370**, 20140183 (2015).

12. Trueta, C. & De-Miguel, F. F. Extrasynaptic exocytosis and its mechanisms: a source of molecules mediating volume transmission in the nervous system. *Front. Physiol.* **3**, 319 (2012).

13. Turing, A. M. The chemical basis of morphogenesis. *Philos. Trans. R. Soc. Lond. B. Biol. Sci.* **237**, 37–72 (1952).

14. Scholes, N. S., Schnoerr, D., Isalan, M. & Stumpf, M. P. H. A comprehensive network atlas reveals that turing patterns are common but not robust. *Cell Syst.* **9**, 243-257.e4 (2019).

15. Miyazawa, S., Okamoto, M. & Kondo, S. Blending of animal colour patterns by hybridization. *Nat. Commun.* **1**, 66 (2010).

16. Josselyn, S. A. & Tonegawa, S. Memory engrams: recalling the past and imagining the future. *Science* **367**, eaaw4325 (2020).

17. Ryan, T. J. & Frankland, P. W. Forgetting as a form of adaptive engram cell plasticity. *Nat. Rev. Neurosci.* **23**, 173–186 (2022).

18. Reinert, S., Hübener, M., Bonhoeffer, T. & Goltstein, P. M. Mouse prefrontal cortex represents learned rules for categorization. *Nature* 1–7 (2021) doi:10.1038/s41586-021-03452-z.

19. Sirota, A. *et al.* Entrainment of neocortical neurons and gamma oscillations by the hippocampal theta rhythm. *Neuron* **60**, 683–97 (2008).

20. Schreiner, T., Doeller, C. F., Jensen, O., Rasch, B. & Staudigl, T. Theta phase-coordinated memory reactivation reoccurs in a slow-oscillatory rhythm during NREM sleep. *Cell Rep.* **25**, 296–301 (2018).

21. Todorova, R. & Zugaro, M. Isolated cortical computations during delta waves support memory consolidation. *Science (80-.).* **366**, 377–381 (2019).

22. Klimesch, W. α-band oscillations, attention, and controlled access to stored information. *Trends Cogn. Sci.* **16**, 606–17 (2012).

23. Terada, S. *et al.* Adaptive stimulus selection for consolidation in the hippocampus. *Nature* **601**, 240–244 (2022).

24. Ngo, H.-V., Fell, J. & Staresina, B. Sleep spindles mediate hippocampal-neocortical coupling during long-duration ripples. *Elife* **9** (2020).

25. Vaz, A. P., Inati, S. K., Brunel, N. & Zaghloul, K. A. Coupled ripple oscillations between the medial temporal lobe and neocortex retrieve human memory. *Science* **363**, 975–978 (2019).

26. Norman, Y. *et al.* Hippocampal sharp-wave ripples linked to visual episodic recollection in humans. *Science* **365** (2019).

27. Leonard, T. K. *et al.* Sharp wave ripples during visual exploration in the primate hippocampus. *J. Neurosci.* **35**, 14771–14782 (2015).

28. Cohn-Sheehy, B. I. *et al.* The hippocampus constructs narrative memories across distant events. *Curr. Biol.* (2021) doi:10.1016/j.cub.2021.09.013.

29. Zheng, J. *et al.* Neurons detect cognitive boundaries to structure episodic memories in humans. *Nat. Neurosci.* **25**, 358–368 (2022).

30. Braitenberg, V. & Schüz, A. Some anatomical comments on the hippocampus. in *Neurobiology of the Hippocampus* (ed. Serfert, W.) 21–37 (Academic Press, 1983).

31. Eichenbaum, H. Hippocampus: cognitive processes and neural representations that underlie declarative memory. *Neuron* **44**, 109–20 (2004).

32. Morris, R. G. M. *et al.* Elements of a neurobiological theory of the hippocampus: the role of activity-dependent synaptic plasticity in memory. *Philos. Trans. R. Soc. Lond. B. Biol. Sci.* **358**, 773–86 (2003).

33. Squire, L. R. & Alvarez, P. Retrograde amnesia and memory consolidation: a neurobiological perspective. *Curr. Opin. Neurobiol.* **5**, 169–77 (1995).

34. McClelland, J. L., McNaughton, B. L. & O'Reilly, R. C. Why there are complementary learning systems in the hippocampus and neocortex: insights from the successes and failures of connectionist models of learning and memory. *Psychol. Rev.* **102**, 419–457 (1995).

35. Frankland, P. W. & Bontempi, B. The organization of recent and remote memories. *Nat. Rev. Neurosci.* **6**, 119–130 (2005).

36. Yu, D., Keene, A. C., Srivatsan, A., Waddell, S. & Davis, R. L. Drosophila DPM neurons form a delayed and branch-specific memory trace after olfactory classical conditioning. *Cell* **123**, 945–957 (2005).

37. Cichon, J. & Gan, W.-B. Branch-specific dendritic Ca(2+) spikes cause persistent synaptic plasticity. *Nature* **520**, 180–5 (2015).

38. Kerlin, A. *et al.* Functional clustering of dendritic activity during decision-making. *Elife* **8** (2019).

39. Schlichting, M. L., Guarino, K. F., Roome, H. E. & Preston, A. R. Developmental differences in memory reactivation relate to encoding and inference in the human brain. *Nat. Hum. Behav.* (2021) doi:10.1038/s41562-021-01206-5.

40. Goshen, I. *et al.* Dynamics of retrieval strategies for remote memories. *Cell* **147**, 678–89 (2011).

41. Monosov, I. E., Haber, S. N., Leuthardt, E. C. & Jezzini, A. Anterior cingulate cortex and the control of dynamic behavior in primates. *Curr. Biol.* **30**, R1442–R1454 (2020).

42. Mansouri, F. A., Freedman, D. J. & Buckley, M. J. Emergence of abstract rules in the primate brain. *Nat. Rev. Neurosci.* (2020) doi:10.1038/s41583-020-0364-5.

43. Hunt, L. T. *et al.* Formalizing planning and information search in naturalistic decision-making. *Nat. Neurosci.* **24**, 1051–1064 (2021).

44. Aggleton, J. P. & O'Mara, S. M. The anterior thalamic nuclei: core components of a tripartite episodic memory system. *Nat. Rev. Neurosci.* (2022) doi:10.1038/s41583-022-00591-8.

45. DF, S. & JS, H. Seasonal hippocampal plasticity in food-storing birds. *Philos. Trans. R. Soc. Lond. B. Biol. Sci.* **365**, 933–943 (2010).

46. Rakic, P. Evolution of the neocortex: a perspective from developmental biology. *Nat. Rev. Neurosci.* **10**, 724–35 (2009).

47. Urbán, N., Blomfield, I. M. & Guillemot, F. Quiescence of adult mammalian neural stem cells: a highly regulated rest. *Neuron* **104**, 834–848 (2019).

48. Denoth-Lippuner, A. & Jessberger, S. Formation and integration of new neurons in the adult hippocampus. *Nat. Rev. Neurosci.* **22**, 223–236 (2021).

49. Kuhn, H. G., Dickinson-Anson, H. & Gage, F. H. Neurogenesis in the dentate gyrus of the adult rat: age-related decrease of neuronal progenitor proliferation. *J. Neurosci.* **16**, 2027–33 (1996).

50. Kandel, P. *et al.* Oleic acid is an endogenous ligand of TLX/NR2E1 that triggers hippocampal neurogenesis. *Proc. Natl. Acad. Sci.* **119** (2022).

51. Spalding, K. L. *et al.* Dynamics of hippocampal neurogenesis in adult humans. *Cell* **153**, 1219–1227 (2013).

52. Tobin, M. K. *et al.* Human hippocampal neurogenesis persists in aged adults and Alzheimer's disease patients. *Cell Stem Cell* **24**, 974-982.e3 (2019).

53. Jurkowski, M. P. *et al.* Beyond the hippocampus and the SVZ: adult neurogenesis throughout the brain. *Front. Cell. Neurosci.* **14**, 576444 (2020).

54. Franjic, D. *et al.* Transcriptomic taxonomy and neurogenic trajectories of adult human, macaque, and pig hippocampal and entorhinal cells. *Neuron* **110**, 452-469.e14 (2022).

55. Krienen, F. M. *et al.* Innovations present in the primate interneuron repertoire. *Nature* **586**, 262–269 (2020).

56. Cossart, R. & Garel, S. Step by step: cells with multiple functions in cortical circuit assembly. *Nat. Rev. Neurosci.* (2022) doi:10.1038/s41583-022-00585-6.

57. Ammothumkandy, A. *et al.* Altered adult neurogenesis and gliogenesis in patients with mesial temporal lobe epilepsy. *Nat. Neurosci.* **25**, 493–503 (2022).

58. Tensaouti, Y., Stephanz, E. P., Yu, T.-S. & Kernie, S. G. ApoE regulates the development of adult newborn hippocampal neurons. *eneuro* **5**, ENEURO.0155-18.2018 (2018).

59. Yu, T.-S. *et al.* Astrocytic ApoE underlies maturation of hippocampal neurons and cognitive recovery after traumatic brain injury in mice. *Commun. Biol.* **4**, 1303 (2021).

60. Loera-Valencia, R. *et al.* High levels of 27-hydroxycholesterol results in synaptic plasticity alterations in the hippocampus. *Sci. Rep.* **11**, 3736 (2021).

61. Jin, W.-N. *et al.* Neuroblast senescence in the aged brain augments natural killer cell cytotoxicity leading to impaired neurogenesis and cognition. *Nat. Neurosci.* **24**, 61–73 (2021).

62. Diotel, N., Lübke, L., Strähle, U. & Rastegar, S. Common and distinct features of adult neurogenesis and regeneration in the telencephalon of zebrafish and mammals. *Front. Neurosci.* **14**, 568930 (2020).

63. Danielson, N. B. *et al.* Distinct contribution of adult-born hippocampal granule cells to context encoding. *Neuron* **90**, 101–12 (2016).

64. Klinzing, J. G., Niethard, N. & Born, J. Mechanisms of systems memory consolidation during sleep. *Nat. Neurosci.* **22**, 1598–1610 (2019).

65. Nägerl, U. V., Köstinger, G., Anderson, J. C., Martin, K. A. C. & Bonhoeffer, T. Protracted synaptogenesis after activity-dependent spinogenesis in hippocampal neurons. *J. Neurosci.* **27**, 8149–56 (2007).

66. Sahay, A. *et al.* Increasing adult hippocampal neurogenesis is sufficient to improve pattern separation. *Nature* **472**, 466–70 (2011).

67. Stone, S. S. D. *et al.* Stimulation of entorhinal cortex promotes adult neurogenesis and facilitates spatial memory. *J. Neurosci.* **31**, 13469–84 (2011).

68. Akers, K. G. *et al.* Hippocampal neurogenesis regulates forgetting during adulthood and infancy. *Science (80-.).* **344**, 598–602 (2014).

69. Anacker, C. & Hen, R. Adult hippocampal neurogenesis and cognitive flexibility-linking memory and mood. *Nat. Rev. Neurosci.* **18**, 335–346 (2017).

70. Sheffield, M. E. & Dombeck, D. A. Dendritic mechanisms of hippocampal place field formation. *Curr. Opin. Neurobiol.* **54**, 1–11 (2019).

71. MEJ, S., MD, A. & DA, D. Increased prevalence of calcium transients across the dendritic arbor during place field formation. *Neuron* **96**, 490-504.e5 (2017).

72. Berry, K. P. & Nedivi, E. Spine dynamics: are they all the same? *Neuron* **96**, 43–55 (2017).

73. Bonetto, G., Belin, D. & Káradóttir, R. T. Myelin: a gatekeeper of activity-dependent circuit plasticity? *Science* **374**, eaba6905 (2021).

74. Yang, S. M., Michel, K., Jokhi, V., Nedivi, E. & Arlotta, P. Neuron class–specific responses govern adaptive myelin remodeling in the neocortex. *Science (80-.).* **370**, eabd2109 (2020).

75. Liu, Y., Foustoukos, G., Crochet, S. & Petersen, C. C. H. Axonal and dendritic morphology of excitatory neurons in layer 2/3 mouse barrel cortex imaged through whole-brain two-photon tomography and registered to a digital brain atlas. *Front. Neuroanat.* **15**, 791015 (2021).

76. Grosmark, A. D. & Buzsáki, G. Diversity in neural firing dynamics supports both rigid and learned hippocampal sequences. *Science* **351**, 1440–3 (2016).

77. Davidson, T. J., Kloosterman, F. & Wilson, M. A. Hippocampal replay of extended experience. *Neuron* **63**, 497–507 (2009).

Chapter 5

Memory in Dendritic Spines

Abstract

We discussed the neuronal basis of memory in Chapter 4. Down to subcellular level, the weight of each synapse probably lies in the shape of each dendritic spine. The neck and head of dendritic spines have different responsibilities. Postsynaptic scaffolds are necessary for a stable synapse. Glutamate goes through the given number of receptors at the synapse, and the electric signal emerges through a dendritic spine into the shaft and the cell body. Despite strong signals within a spine, the activation can remain local, and the weights are updated competitively with activity, during sharp-wave ripples. Forward and reverse replays accumulate synaptic weights, update the grouping of neurons and promote storage of new memory.

Keywords

Dendrites, Dendritic spines, Long-term potentiation (LTP), α-Amino-3-hydroxy-5-methyl-4-isoxazolepropionate (AMPA)-type glutamate receptors, Synaptic weights, Electrical resistance, Local computation, Replay, Sharp-wave ripples (SWRs)

5.1 Spiny Neurons

Individual neurons do a lot of computing[1-5]. The idea of dendritic spines (protrusions on dendrites that most axons form synapses

Neuroscience for Artificial Intelligence
Huijue Jia
Copyright © 2023 Jenny Stanford Publishing Pte. Ltd.
ISBN 978-981-4968-78-2 (Hardcover), 978-1-003-41098-0 (eBook)
www.jennystanford.com

with, Figs. 5.1 and 1.9) being important for learning and memory goes back over a hundred years ago, as Santiago Ramón y Cajal saw them in stained images of neurons[6]. As summarized by Dr. Francis Crick, the neck of a dendritic spine is likely a parameter for defining synaptic weight, subject to both immediate (e.g., due to actin contraction in < 10 ms) and long-term modulation[7-10] (Figs. 5.1, 5.2, 5.3, 5.4 and 5.5).

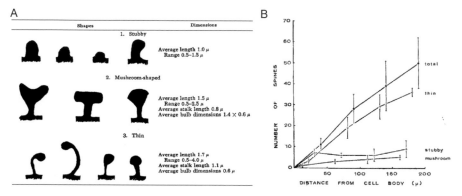

Figure 5.1 (A) Shapes and length of dendritic spines. These would be all over the branches of pyramidal (Fig. 4.1) and spiny stellate cells*[,1, 11]. (B) The distribution of dendritic spines in Golgi preparations of apical dendrites from pyramidal cells located in layer III. The number of spines in 50 μm lengths is given. Each point on the graph is the mean of ten values and the extent of the vertical line above and below each point represents + or - one standard deviation. The spines change during development and in adulthood[12].

Credit: Part A from Table 1 of ref. 13, which was also Fig. 2 of ref. 7. Part B from Fig. 2 of ref. 13.

Inhibitory synapses are larger and tend to be on dendritic shafts (Fig. 5.4, Section 1.3), and some excitatory axons also synapse on shafts instead of spines[14].

The majority of dendritic spines are thin and close to the ends of branches far from the cell body (soma, where the nucleus lies, Fig. 5.1B) where the dendritic branches are also slenderer and have a higher electric resistance (Fig. 5.2). So the thin spines would have an electric resistance of hundreds of MΩ and often work in isolation both biochemically and electrically (Figs. 5.2 and 5.3)[15-17]. Being far from the cell body (Fig. 5.1B) means that even when signals do emerge from these spines, they will arrive at the cell body milliseconds later[18], when they could

*The term "stellate cell" is inconsistently used in the literature and more consensus is needed.

be overwhelmed (or facilitated) by other signals from other branches of the neuron.

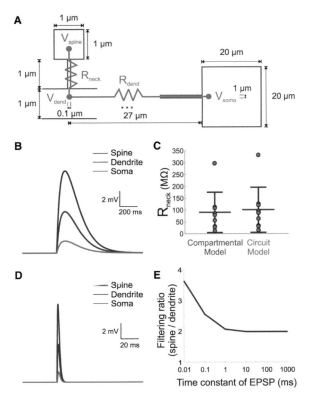

Figure 5.2 Estimation of spine neck resistance. (A) A compartment model (purple) and an electrical circuit model (green) of a cultured mouse hippocampal neuron were built to determine spine neck resistance. (B) To estimate the spine neck resistance, its voltage profiles were simulated with the compartmental model and average dendritic resistance, 108 MΩ estimated from ref. 15. The neck resistance was 95 MΩ. (C) Multiple simulations of (B) with the range of dendritic resistance (63–153 MΩ, R_{dend}) resulted in the range of neck resistance from 64 MΩ to 146 MΩ (purple). Alternatively, the range of neck resistance was calculated by the electrical circuit model and produced similar results from 63 MΩ to 153 MΩ (green). (D) With fast excitatory postsynaptic potentials (EPSP) of 1ms time constant, the compartmental model was simulated like (B) and its filtering ratio was acquired, similar to that acquired by slower EPSP. (E) The simulation of (D) was performed with various kinetics of EPSP (τ = 0.01, 0.1, 1, 10, 100, 1000 ms). The result indicates that the filtering ratio estimated by slow uncaging EPSP is still valid in faster event around EPSP 1–10 ms time constant, which is more common in physiology. Error bars: SD.

Credit: Fig. 4 of ref. 16.

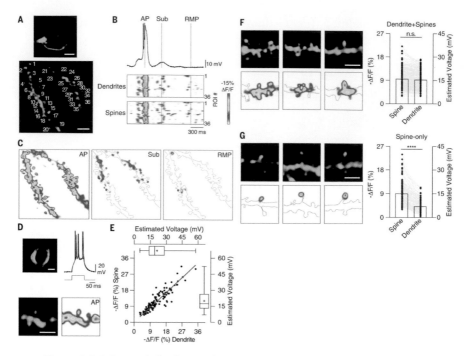

Figure 5.3 Spine and dendritic voltage dynamics in vivo during spontaneous activity. (A) Top: In vivo two-photon imaging and somatic whole-cell of neurons expressing postsynaptic Accelerated Sensor of Action Potentials (postASAP). Red lines: pipette outline. Bottom: imaged dendrites (43 μm from center of the image to cell body) of patched cell, scales = 5 μm. (B) Top: somatic electrical recording of neuron in (A). AP: train of action potentials; Sub: subthreshold depolarization; RMP: resting membrane potential. Bottom, simultaneous fluorescence changes of numbered spines and adjacent dendrites in (A). (C) Representative image with peak fluorescence changes in dendrites and spines during three conditions in (B). (D) Depolarization during APs, generated by three 100 ms current pulses (300 pA). Top: somatic imaging and electrophysiological recording. Bottom: representative fluorescence changes in dendrite [average 3 trials; spine at 48 μm from cell body, scale = 5 μm; color scale same as (B)]. (E) Peak spine and dendrite fluorescence changes during AP trains (n = 125 spines, 37 dendrite segments, 5 cells, 4 animals; linear regression: y = 0.93x, $R2$ = 0.823, p < 0.0001). (F) Left: examples Dendrite+Spines patterns (average 10 events). Right: peak fluorescence changes in spine heads and adjacent dendrites (n = 221 spines, 90 dendritic segments, 13 cells, 7 animals). (G) Same as (F) for Spine-only pattern (n = 116 spines, 90 dendritic segments, 13 cells, 7 animals).

Credit: Fig. 2 of ref. 17.

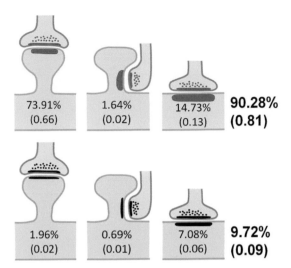

Figure 5.4 Schematic representation of the distribution of asymmetric (excitatory, neurotransmitter glutamate) synapses (*green*) and symmetric (inhibitory, neurotransmitter gamma aminobutyric acid (GABA)) synapses (*red*) on spines and dendritic shafts. A total of 6184 synapses in all six layers of the somatosensory cortex representing the hindlimb of juvenile rats (postnatal day 14, three males) were serial sectioned using focused ion beam milling and scanning electron microscopy (FIB/SEM). Synapses on spines have been sub-classified into those that are established on the head of the spine and those that are established on the neck. Percentages represent the average of the six cortical layers. Values between parentheses represent the density of each type of synapse in the neuropil, in synapses per μm³.

Credit: Fig. 3 of ref. 19.

As mature spines, the thin spines and mushroom-shaped spines (Fig. 5.1) contain in their heads the PSD-95 (Postsynaptic Density protein-95) scaffold proteins that binds both the NMDA (N-methyl-D-aspartate) receptor and the AMPA (α-amino-3-hydroxy-5-methyl-4-isoxazole propionic acid)-type glutamate receptor[20] (Section 5.2). A recent study appeared to have included thin spines and stubby spines among filopodia (head/diameter ratio < 1.3 and length/head diameter ratio >3), reporting 30% of all dendritic protrusions as filopodia (adult mice L5 primary visual cortex), among which 25% only showed NMDA and no AMPA.[21] The results are still consistent with mature spines having more AMPA than NMDA[20]. The amount of PSD-95 correlated with the volume of a spine[14].

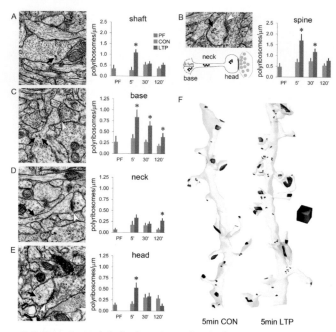

Figure 5.5 Transient global elevation of polyribosomes in dendritic shafts and spines was followed by sustained elevation of polyribosomes (multiple ribosomes, ~30 nm each, lined up on an mRNA) in spine bases and necks during long-term potentiation (LTP). (A) Left: Electron microscopy (EM) image of a polyribosome (arrow) in a dendritic shaft from the 120 min LTP condition. Right: There were more polyribosomes in dendritic shafts in the 5 min LTP condition ($F_{(1, 37)}$ = 14.12, P = 0.00059; LTP × experiment interaction $F_{(1,35)}$ = 12.23, P = 0.0013). (B) Top left: EM of a polyribosome (arrow) in the head of a dendritic spine receiving a synapse (arrowhead) from the 120 min LTP condition. Bottom left: Diagram of locations of polyribosomes (black) within a dendritic spine. The PSD (postsynaptic density, red) and presynaptic vesicles (blue) are also shown. Right: There were more polyribosomes in dendritic spines with LTP at 5 min ($F_{(1, 37)}$ = 9.20, P = 0.0044) and 30 min ($F_{(1, 49)}$ = 4.77, P = 0.034). (C–E) Left: EMs of polyribosomes (arrows) in a spine base (C), neck (D), and head (E) from the 120 min LTP (C) and 120 min control (d–e) conditions; arrowheads indicate synapses. Right: There were more polyribosomes in the base of dendritic spines (C) 5 min ($F_{(1, 37)}$ = 6.41, P = 0.016), 30 min ($F_{(1, 49)}$ = 10.49, P = 0.0022), and 120 min ($F_{(1, 44)}$ = 4.39, P = 0.042) after LTP induction. There were no differences among the three LTP groups ($F_{(2, 63)}$ = 3.10, P = 0.052). There were more polyribosomes in spine necks (D) at 120 min ($F_{(1, 44)}$ = 12.25, P = 0.0011) and in spine heads (E) at 5 min ($F_{(1, 37)}$ = 8.78, P = 0.0053). (F) Reconstructed dendrites from the 5 min experiment (left, control; right, LTP) showing polyribosomes (black) and synapses (red). Effects at $P < 0.05$: *control vs LTP; # control vs perfused. Scale in *A–E* = 250 nm, blue cube in *F* = 500 nm/side.

Credit: Fig. 2 of ref. 22.

The stubby spines are immature, can recruit more than one ribosomes to make proteins[22] (Figs. 5.1 and 5.5), and many would not survive a night's sleep as they compete with each other (Section 5.2).

There are also filopodia (slender feet)-like structures that are even more tentative than the stubby spines, constituting 2–3% of the spines in a mature brain but much more in infants[20, 23]. Some of the filopodia might grow to become dendritic shafts instead of spines[22]. After stochastic initiation of filopodia from a dendritic shaft, an actin-driven filopodia-like structure has an average lifetime of minutes to hours[20], as it grows towards a nearby axon[23]. Proteins that regulate the actin dynamics and proteins that bind to positive and negative membrane curvature underlie the filopodia and dendritic spine formation[24], favoring filopodia generation at thin ends of dendrites[23].

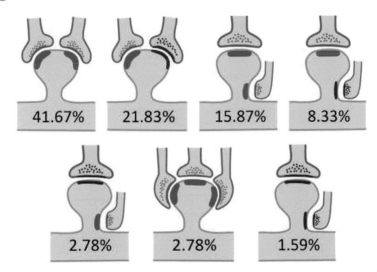

Figure 5.6 Schematic representation of the different locations of multiple synapses on the head and neck of spines in the rat somatosensory cortex[19]. The proportions could be different for a different region of the cortex (e.g., ref. 28). Excitatory (asymmetric; neurotransmitter glutamate) synapses have been represented in *green* and inhibitory (symmetric; neurotransmitter GABA) synapses in *red*. 5.57% (254) of the 4558 dendritic spines (Fig. 5.2) had two or three synapses, and the breakdown is shown here. Percentages indicate the relative frequency of each case with respect to the total number of spines establishing multiple synapses. Other combinations were rarely found (about 5% of all cases) and have not been represented.

Credit: Fig. 4 of ref. 19.

With the established framework of neighboring neurons[11], to shift a connection along a dendrite towards the cell body, for a faster and likely shorter and more robust spine (Fig. 5.1; and to potentially store a more general rule, Chapter 4), might involve switching to another presynaptic partner neuron.

Synaptic weights differ by orders of magnitude (e.g., Fig. 5.1). More complicated than the average synapse size mentioned in Section 1.3, the distribution of dendritic spine head volume was observed to be a binary mixture of two log-normal distributions for pyramidal cells in layer II/III of the mouse primary visual cortex[25]; The distribution of synapse size was also binary for hippocampal CA1 pyramidal cells[26]. To change one or two large synapses may be enough to modulate the optimal orientation of the signal that is detected through the dendrites by the neuron (Fig. 4.3), which can lead to an action potential involving the entire cell[27] (Figs. 5.2 and 5.3).

Whereas the majority of spines have no more than one synapse, more complex combinations have been occasionally found in spines with two or more synapses, both on the spine head, or one on the head and the other on the neck (Fig. 5.6).

5.2 Local Spine Dynamics

5.2.1 Memory Decay Down to Individual Spines

Single-cell recordings became available only in recent years, facilitating our understanding of memory engrams (Chapter 4). However, the physical basis of memory would ultimately come down to individual synapses. Individual dendritic spines are tentatively generated every day (Section 5.1), and likely compete for resources, so that some stubby spines are demolished while the total number of spines and the large spines are relatively stable every day[29-31] (Fig. 5.7; more on sleep in Chapter 6, which likely also contribute to the differential accumulation of resources in some neurons more than others).

5.2.2 New and Leaky

Ca^{2+} transients mediated by NMDA-type glutamate receptors (Fig. 4.3) were significantly lower in new spines[32]. The Ca^{2+} could diffuse from the new spine into the dendrite[32], which would otherwise accumulate in the individual spine and facilitate spine enlargement if the neck is narrow[33] (Fig. 5.8, reminiscent of Fig. 5.3).

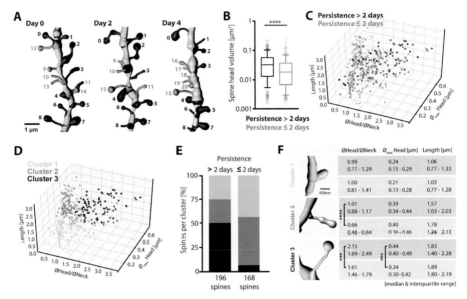

Figure 5.7 Structure-dynamics relationship of hippocampal spines. (A) 3D reconstruction of a dendrite imaged on days 0, 2 and 4 using two-photon super-resolution stimulated emission depletion (2P-STED) microscopy. Spines persisting for more than 2 days (#0–8, blue), and 2 days or less (#9–20, salmon) are illustrated. (B) Spine head volumes measured on reconstructed dendrites ($p < 0.0001$, Mann-Whitney test; $n = 14$ dendrites, 3 mice; box plot shows median and 10, 25, 75 and 90th percentiles). (C, D) 3D morphology plots visualizing the populations of spines observed persistent for more than 2 days and 2 days or less (C), and their affiliation to identified clusters 1, 2 and 3 (D). Plotted are, the ratio of mean head to neck diameters (ØHead/ØNeck), spine length and maximum head diameter (Ømax Head). (E) Quantification of spine proportions within identified clusters, distinguishing spines of different persistence (>2 days versus ≤2 days). (F) Table summarizing the morphological parameters utilized for cluster analysis: ØHead/ØNeck, Ømax Head and length of spines, for spines that persist for >2 days (blue) and ≤2 days (salmon). Data are represented as median and interquartile range (25th–75th percentile). Significant differences are marked by asterisks (***$p < 0.001$, ****$p < 0.0001$, unpaired t-test or Mann-Whitney test; $n = 14$ dendrites, 3 mice). The larger heads have likely better survived remodeling of synapses during sleep (Chapter 6).

Credit: Fig. 4 of ref. 29.

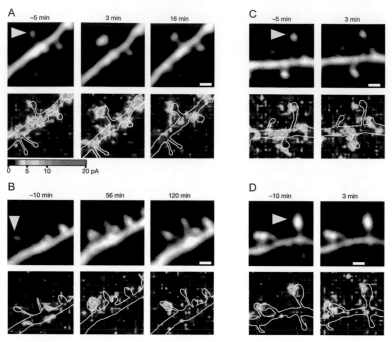

Figure 5.8 Colocalization of enlargement of spine heads and potentiation of AMPA-receptor-mediated currents. Glutamate uncaging was evoked by two-photon photolysis of 4-methoxy-7-nitroindolinyl (MNI)-caged glutamate at individual spines of cultured rat hippocampal slices. (A, B) Examples of small spines that showed transient (A) or long-lasting (B) enlargement (upper panels) and potentiation of AMPA currents (lower panels). Three-dimensional mapping of AMPA currents was performed in neurons that were clamped in the perforated-patch mode. The amplitude of AMPA currents is pseudocolor coded and stacked along the z axis by the maximum-intensity method. The neurons were depolarized to 0 mV and the small spines, indicated by the arrowheads, were stimulated by two-photon uncaging of MNI-glutamate at 2 Hz between times 0 and 60 s. White lines in the lower panels indicate contours of dendrites. Scale bars, 1 μm. c, d, Examples of small (c) and large (d) spines whose heads failed to enlarge in response to paired stimulation.

Credit: Fig. 3 of ref. 8.

Regardless of spine neck (Figs. 5.1 and 5.7), the number of AMPA-type glutamate receptors increased with increasing volume of a dendritic spine[32] (Fig. 5.9), or with the area of postsynaptic density (PSD), dependent on actin polymerization[8]. The area of PSD (the number of receptors) and the resulting current correlated with vesicle release probability at the synapse[18].

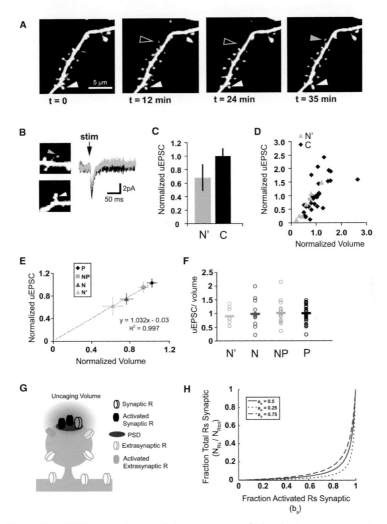

Figure 5.9 AMPA receptor-mediated currents of developing spines mature coincident with increase in spine volume. (A) Dendrites of GFP-transfected hippocampal pyramidal neurons in cultured rat hippocampal slices were imaged every 10–12 min in artificial cerebrospinal fluid (ACSF) at 35°C. Spines were classified into two groups: control (C; present at all time points; e.g., white arrowhead), and early new (N'; appearing after the first time point; e.g., green arrowhead). Early new spines were all less than 50 min old. (B) After time-lapse imaging, the cells were patched and whole-cell currents were recorded at the soma. Shown are current recordings in response to glutamate uncaging, evoked by two-photon photolysis of MNI-caged glutamate, at the control (black trace) and early new (green trace) spines

(Continued)

Figure 5.9 (*Continued*)

identified in (A). Each trace is the average of five to seven trials. Vertical black arrow (stim) marks the time of the stimulus. pA, picoampere. (C) AMPA current amplitudes (mean ± SEM) from early new (N'; $n = 7$) and control (C; $n = 31$) spines normalized to the mean of all (≥ 3) C currents from the same dendrite ($N = 7$ cells). uEPSCs, uncaging-evoked excitatory currents. AMPA current amplitudes of early new spines are smaller than those of control spines ($p < 0.1$; one-tailed t-test). (D) Normalized AMPA current amplitudes plotted against normalized volumes for control (C; black diamonds), and early new (N'; green triangles) spines. AMPA current amplitudes and spine volumes of control spines are highly correlated ($r = 0.65$; $p < 0.01$; 31 control spines). Data from early new spines appear similarly correlated ($r = 0.98$; $p < 0.01$; $n = 7$). (E) Mean normalized AMPA current amplitudes plotted against mean normalized volumes for early new (N'; green triangle), new (N; red triangle), new persistent (NP; blue square), and persistent (P; black diamond) spines. Across developmental time, mean AMPA current amplitudes increase linearly with increase in spine volume ($R2 = 0.999$). Error bars represent SEM. (F) Summary of normalized AMPA current amplitude to volume relationships for individual early new (N'; $n = 8$), new (N; $n = 12$), new persistent (NP; $n = 15$), and persistent (P; $n = 37$) spines. Horizontal bars represent mean values. (G) Schematic of the experimental configuration, showing the activation of synaptic and extrasynaptic receptors that are within the cloud of uncaged glutamate. The intersection of this cloud with the spine head defines the photoactivated spine head area (aA). As detailed in panel H, extrasynaptic AMPA receptors make a negligible contribution[32]. (H) Fraction of total receptors that are synaptic versus the fractional contribution of synaptic receptors to the uEPSC. Plots correspond to different values for the fraction of the photoactivated spine head area (aA = 0.25, dotted line; aA = 0.5, continuous line; aA = 0.75, broken line). The fraction of the total surface area on spines, RSH/tot = 0.0595, was derived from EM reconstructions. If synaptic receptors contribute less than 80% to the uEPSC (bs < 0.8), then less than 10% of all receptors would be synaptic (NRs/NRtot < 0.1), which would contradict microscopic evidence.

Credit: Fig. 2 of ref. 32.

5.2.3   Thin and Learning Fast

Computationally, the prevalence of thin spines[25] (Figs. 5.1 and 5.2; Fig. 3.6 in newborn guinea pig) represent uncertainty that can be quickly updated by a single input of new information (Fig. 5.8; a likely explanation for the binary log-normal distribution of synaptic weight), which is an important setup for Bayesian inference[34]. Nearby spines could also be affected by the resource package delivered[8] (Figs. 5.3 and 5.8).

5.3 Memory Replays at Synapses

Studies in rats or mice typically involve running on a linear track, but the "velocity" of the memory replay compared to the animals' running speed is likely irrelevant. I would rather ask how many synapses are involved (see also sensory input in Chapter 2). Spontaneous replays that span dozens to hundreds of milliseconds (ms)[35], occur due to sharp-wave ripples during slow-wave sleep and when awake[36, 37] (Chapter 6). A typical synapse takes 5–10 ms[11, 38, 39]; Ca^{2+} diffusion from a single spine into the dendritic shaft can last 15 ms[33], which might activate nearby synapses (Figs. 5.2 and 5.3) with other cells which then have their own currents and calcium. This could also include parallel detours in the brain that come back to the same neuron.

A study in mice showed that stimulation of the anterior cingulate cortex (ACC)-to-hippocampal CA1 axonal terminals (directed search, Fig. 4.11), but not the cell body (soma) of the hippocampal CA1 neurons, was sufficient to retrieve recently learned fear memories[40]. The ACC subregion of the prefrontal cortex is important for memory consolidation (Section 4.4). Actin destabilization in ACC during the second day of sleep was recently shown to erase fear memory of electric foot shock[41], consistent with the time line of dendritic spine growth in ACC pyramidal neurons (e.g., L2/3[42]) one synapse upstream of the hippocampal CA1 spines. The medial entorhinal cortex (MEC) can also generate replays[43].

With all the necessary information in dendrites, fast replays do not have to activate every synapse, rather, the first ones within each cluster to get depolarized might be sufficiently representative (more on replays and preplays (planning) in Chapters 6 and 9). So the activation could also be sparse and frugal.

5.4 Sharp-Wave Ripples—Weights of Dendritic Spines in Action

Electric waves of different frequency can be detected when awake and during sleep (e.g., Fig. 5.10, Chapter 6). It is currently believed that sharp-wave ripples (SWRs, ~200 Hz) drive replays or preplays[36] (Section 5.3, Chapter 9), and are particularly relevant for memory and forgetting[44]. They can appear sporadically on top of the slower waves such as delta waves[45] (Fig. 6.5, synchronization by interneurons), theta waves[46] (Section 6.5).

But we discussed in Section 5.3 that replays likely represent clusters of synaptic activation, and the frequency and voltage of SWRs are also consistent with actual activity in individual synapses, i.e., dendritic spines (Figs. 5.2 and 5.3). As we mentioned in Section 4.4, the slower waves likely represent a broader search (involving interneurons) that modulate the probability of synaptic activities.

According to mice experiments, electrical stimulation of the hippocampal CA1 excited neurons in the ACC (anterior cingulate cortex) with a mean latency of 23 ms (5–60 ms range)[47]. The mean latency between the peak of hippocampal sharp-wave ripple activity and the peak firing of ACC type I neurons was 12 ms (6–32 ms range; ACC type I neurons were defined as neurons that responded within 100 ms)[47]. The projection was mostly monosynaptic from ACC to the hippocampus[48]. About one quarter of CA1 ripple events led to activation of ACC type I neurons, which were recruited after individual ripple events[47]. Almost all ACC neurons increased their activity just before hippocampal ripple activity during slow-wave sleep (Chapter 6, Fig. 6.5); and hippocampal CA1 ripple activity increased during the depolarizing instead of the hyperpolarizing phase of cortical slow-wave oscillation[47].

When human volunteers were shown pictures of landscapes or buildings before a nap, a 500–1200 ms late replay associated with remembering and a 100–500 ms early replay associated with forgetting[49] (Fig. 5.10). The pictures with an early replay were possibly regarded as familiar (Section 7.8, phase precession). Monosynaptic input from the entorhinal cortex L3 arrives at the peak of hippocampal CA1 θ waves and has been suggested to drive reverse replay, while hippocampal CA3 input arrives at the trough of CA1 θ waves and drives forward replay[50] (Section 5.5). The longer duration of ripples likely correspond to more features remembered[51].

Simulations from another study showed that the number of postsynaptic bursts per spike modulates synaptic weight[39], so they are working on the individual dendritic spines which together gain resources in the region. Sleep (both NREM and REM sleep, Chapter 6) looks like good opportunities to update the cortex-hippocampal hash function (Fig. 4.11). Cerebellum Purkinje cells also have dendritic spines, and increased presynaptic spikes per second after visual stimulation indicated reinforcement learning[52, 53].

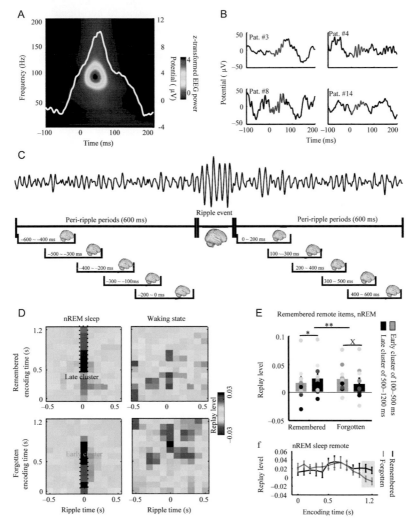

Figure 5.10 Ripple-triggered replay. 12 epilepsy patients were implanted with electro-encephalogram (EEG) electrodes for diagnostic purpose. 80 pictures of landscapes or buildings were learned before a 1 h nap, followed by another 80 pictures after the nap to make the first set "remote." (A) Analysis of replay during ripple events and peri-ripple periods: schematic overview. Top: Example of intracranial EEG activity from the hippocampus (high pass filtered at 40 Hz). Below: time windows used to extract activity during

(Continued)

Figure 5.10 (*Continued*)

peri-ripple periods and ripple events. (B) Time resolved replay of gamma activity locked to ripples during nREM sleep and waking state for later remembered and forgotten remote items. Ripple time 0 corresponds to ripple events and ripple time before and after 0 correspond to peri-ripple period. The early (100–500 ms) and the late cluster (500–1200 ms) correspond to the stimulus-specific clusters, i.e., higher correlations between encoding of one and retrieval of the same item as compared to encoding of one and retrieval of a different item in the gamma range. (C) Ripple-triggered replay of early and late encoding activity differentially affects memory: Interaction between replay levels of remembered and forgotten items and gamma band activity from the early (100–500 ms) and the late (500–1200 ms) encoding clusters. (D) Direct comparisons of replay levels during ripple events between later remembered and later forgotten remote items. Each colorful dot indicates one participant. Same colors indicate data from the same participant. Error bars, standard error of the mean; $*p < 0.05$ (paired t-test); $**p < 0.01$ (2×2 repeated measures ANOVA); X, $p = 0.071$ (paired t-test); time windows showing higher replay levels for remembered vs. forgotten items during ripple events are indicated by gray background.

Credit: Fig. 4c–e of ref. 49.

5.5 Gated Storage of New Details

Coordinated by other cells such as interneurons, it was shown over 20 years ago that synchronous firing of >10% of the rat CA3 cells within a 100 ms window was required to exert a detectable influence on the network oscillation of CA1 pyramidal cells, whereas single pyramidal cells in CA3 can drive ripples in CA1[52]. CA3 to CA1 axons (Schaffer collaterals) assigned to cued experience (sparse encoding, see also CCK basket cells in Section 6.3) no longer participated in spontaneous SWRs, while these that were active during such spontaneous ripples were also suppressed when the cued experience was being replayed as memory[55]. Fifty percent of pyramidal neurons in mice CA1 contain an axon originating from a basal dendrite, which was recently shown to allow recruitment into ripple oscillations[56], while an axon originating from the cell body would remain suppressed by inhibition targeting the cell body (soma, more on interneurons in Section 6.3).

After multiple laps in a new place, hippocampal CA1 place cells have been recruited by conjunctive input from the entorhinal cortex and hippocampal CA3 cells[1, 57, 58], which replay during rest (Fig. 5.11). A recent study simulated spontaneous replays in CA3 using a network of 8000 excitatory pyramidal cells and 150 inhibitory interneurons, and detected replay ensembles as recurring localized sequential activity using a convolutional nonnegative matrix factorization (convNMF) algorithm[55]. Hebbian learning[59]—"Cells that fire together, wire together"—likely require interneurons (which, besides having multiple axons, can also be connected by gap junctions) that output to multiple pyramidal cells (Section 6.3).

As we might guess, the probability for each step of memory formation is not very high (Figs. 5.12 and 5.13)[61], more practice, emotion, or reward (reinforcement learning, Chapter 9) is needed to increase the probability. These also mean that things are not catastrophically overwritten as we age, which is a problem for continual learning algorithms.

According to simulation using a neural network, the mono-synaptic pathway from entorhinal cortex directly to hippocampal CA1 (far from the cell body, i.e., soma) was able to support statistical learning (update the weights); while the trisynaptic pathway from entorhinal cortex (e.g., MEC layer II stellate cells, Section 7.2), dentate gyrus, and CA3 to CA1 (Schaffer collaterals that synapse on apical dendrites of CA1 pyramidal neurons, close to the cell body[64], Fig. 5.14) learned individual episodes[65]. Such diverging pathways are likely common in the nervous system, ensuring immediate fast response and at the same time allowing important new information to be recorded, to prepare for the future.

Note that adult neurogenesis in mice and rats takes place in the hippocampal dentate gyrus (Figs. 4.13 and 5.14), with each 2-week old dentate gyrus granule cells reaching 11–15 pyramidal cells in CA3[66]. Unlike mice, new neurons (including interneurons[67]) likely take much longer to mature in primates, at least 6 months in macaque monkeys[68]. The new neurons likely contain information for both time and space (Fig. 5.14, Chapter 7). Besides episodic memory, the lateral entorhinal cortex (LEC) also contains information on reward (dopamine from the ventral tegmental area and substantia nigra)[69], which would be important for reinforcement learning (Chapter 9).

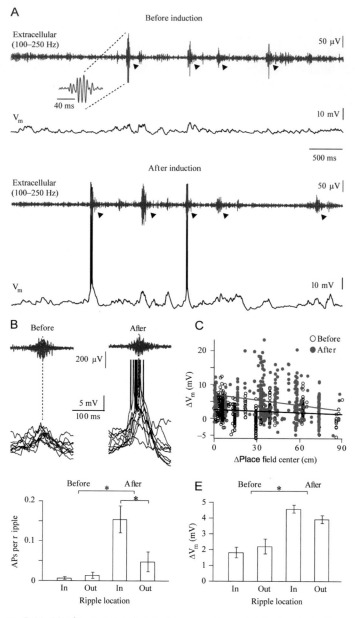

Figure 5.11 Ripple-associated membrane voltage (V_m) depolarization and action potential (AP) output increase after induction of a new CA1 place field. (A) Extracellular ripple (LFP filtered between 100 and 250 Hz) and intracellular V_m recording before (top) and after (bottom) place field

(Continued)

Figure 5.11 (*Continued*)

induction in a CA1 pyramidal neuron. Animal was stationary during the period shown. Arrowheads indicate ripples detected by the algorithm[57]. (C) Relationships between the ripple-associated subthreshold V_m changes (ΔV_m) and the locations of the ripple relative to the place field center (at 0) before (black open circles) and after (gray filled circles) place field formation. Data are taken from the cell shown in (A). Each circle represents one ripple (before, N = 230 ripples; after, N = 340 ripples); the lines show linear fits. Ripple locations before and after the place field center were pooled. (D) Number of APs per ripple for events inside and outside the neuron's place field (mean ± s.e.m., N = 16 neurons, two-tailed unpaired t-test, before versus after $P = 4.0 \times 10^{-4}$, after inside versus after outside P = 0.017). (E) ΔV_m for ripples inside and outside the neuron's place field (mean ± s.e.m., N = 16 neurons, unpaired two-tailed t-test, before versus after $P = 1.0 \times 10^{-4}$).

Credit: Fig. 8 of ref. 57.

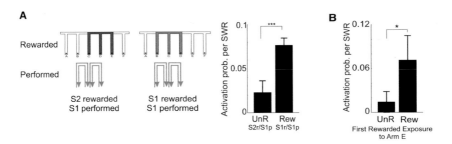

Figure 5.12 Comparison of sharp-wave ripple (SWR) activity between rewarded and unrewarded conditions. (A) Activation probability for the same behavioral sequence with and without rewards (liquid chocolate). Left: schematic of sequence rewarded (highlighted on track, S1 in blue and S2 in red) and sequence performed (arrows below track) for data shown at right. Right: activation probability per wSWR when animals accurately performed S1 and reward was omitted (S2 rewarded, S2r/S1p) or delivered (S1 rewarded, S1r/S1p). Male Long-Evans rats were handled and food deprived to 85%–90% of baseline weight, and pretrained on a linear track with rewards at the ends. The animals were then implanted with a microdrive array containing 16 independently movable tetrodes targeting CA3 (−3.6 mm AP; 3.4 mm L), which were gradually moved down from CA1 to CA3 in 7–10 days. (B) Activation probability per wSWR during the first rewarded exposure to S2, when reward was unexpected based on the previous history of rewards, and unrewarded trials in all arms. Only times when animals were stopped at the well were included. ***indicates $p < 10-5$. Error bars represent standard errors.

Credit: Fig. 6 of ref. 60.

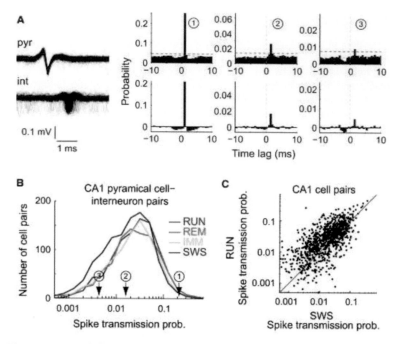

Figure 5.13 Probability distributions of spike transmission in different brain states. (A) Monosynaptic drive of a putative interneuron by a pyramidal cell. Left: superimposed filtered waveforms (800 Hz to 5 kHz) of a pyramidal cell (pyr) and an interneuron (int) triggered by spiking of a pyramidal cell. The two neurons were recorded from different silicon probe shanks. Right: three example cross-correlograms showing short-latency, putative monosynaptic interactions between CA1 pyramidal-interneuron pairs (recorded from two different electrodes). The first example corresponds to the left filtered waveforms. Dashed red lines indicate 0.1% and 99.9% global confidence intervals estimated by spike jittering at a uniform interval of [−5, 5] ms[62]; blue, mean. Note the different magnitude probability scales. Bottom row: shuffling corrected histograms of the same neuron pairs. (B) Distribution of spike transmission probability values (note log scale) between CA1 pyramidal cells and putative interneurons in different brain states. Circled numbers indicate the probability values shown in (A). (C) Comparison of spike transmission probability between RUN and SWS. Note larger values during RUN. Each dot represents a single-cell pair. (D) Spike transmission probability between principal cells and putative interneurons in the CA1, CA3 regions and entorhinal cortex (EC; neuron pairs from EC layers were combined) in different brain states. Median, lower, and upper quartiles are shown. Brackets indicate significant differences ($p < 0.05$, Kruskal-Wallis ANOVA, followed by Tukey's honestly significant difference test).

Credit: Fig. 7 of ref. 63.

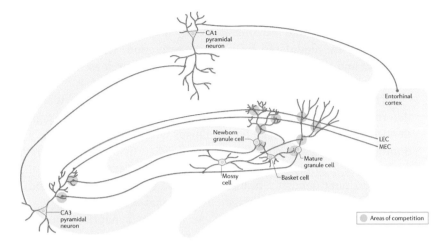

Figure 5.14 Circuit function of newborn granule cells in the adult hippocampus. Newborn granule cells in the dentate gyrus (DG) receive stronger input from the lateral entorhinal cortex (LEC) than from axons arising from the medial entorhinal cortex (MEC). Immature granule cells may affect DG circuit activity by directly exciting mature granule cells or by indirectly inhibiting them via activation of inhibitory interneurons (such as parvalbumin-positive basket cells) or hilar mossy cells. With maturation, granule cells receive strong perisomatic inhibition and form large boutons onto CA3 pyramidal cells, thus modulating output onto area CA3 and area CA1, from where information flows back to the entorhinal cortex and other association cortices. The exact effects of newborn neurons at distinct maturational stages on activity and information flow of the DG circuit and its output into area CA3 will need to be further analyzed by in vitro and in vivo imaging and electrophysiological recordings. Blue shaded areas indicate sites where new neurons may have to compete with pre-existing cells for presynaptic and postsynaptic partners.

Credit: Fig. 4a of ref. 70.

5.6 Summary

Dr. Francis Crick has probably got it right again, by stating in 1982 that synaptic weights lie in the neck of each dendritic spine[7]. The head of each spine contains the proteins that make a stable synapse and determines the maximum current. Most of the mature spines have a long neck (>1μm) and a small head. Their weights are competitively increased or decreased during sharp-wave ripples that probably represent activity in individual synapses

(Chapter 6 on sleep), consistent with Hebbian learning, but with physical transfer and accumulation of materials. A small spine is more easily modulated by a single activation, which enlarges the head and then the neck—So a high probability does not easily go much higher in Bayesian learning.

The shape and distance of dendritic branches would determine the computation after electric signals emerge through dendritic spines[17, 71, 72]. The few synapses with large weights likely dominate replays and also more easily lead to firing of the entire neuronal cell. Filopodia are stochastically generated and can grow into new dendritic branches if there is a nearby axon and if the logistic supply is encouraged by neuronal activity.

Forward and reverse replays accumulate synaptic weights, update the grouping of pyramidal neurons and interneurons if necessary, and promote storage of new memory, including incorporation of new neurons.

Questions

1. With computation within dendritic spines and shafts, how would a single pyramidal neuron correspond to a neural network?

2. How do you think different animals and different individuals might differ in the shape, electric resistance and molecular composition of neurons, and how would these influence neuronal computation?

3. We mentioned in Section 1.3 that available membrane surface around an axon predicts a synapse[73]. Would it be a good idea to initially occupy the surface with an interneuron, and only release the surface for dendrites of pyramidal neurons after an activation? Will existing pyramidal neurons also have to help cover the surface, before a new pyramidal neuron emerges?

References

1. Megías, M., Emri, Z., Freund, T. F. & Gulyás, A. I. Total number and distribution of inhibitory and excitatory synapses on hippocampal CA1 pyramidal cells. *Neuroscience* **102**, 527–540 (2001).

2. Losonczy, A. & Magee, J. C. Integrative properties of radial oblique dendrites in hippocampal CA1 pyramidal neurons. *Neuron* **50**, 291–307 (2006).

3. Poirazi, P. & Papoutsi, A. Illuminating dendritic function with computational models. *Nat. Rev. Neurosci.* **21**, 303–321 (2020).

4. Li, S. *et al.* Dendritic computations captured by an effective point neuron model. *Proc. Natl. Acad. Sci. U. S. A.* **116**, 15244–15252 (2019).

5. Otor, Y. *et al.* Dynamic compartmental computations in tuft dendrites of layer 5 neurons during motor behavior. *Science (80-.).* **376**, 267–275 (2022).

6. Ramón y Cajal, S. *Comparative Study of the Sensory Areas of the Human Cortex.* (Norwood Press, 1899).

7. Crick, F. Do dendritic spines twitch? *Trends Neurosci.* **5**, 44–46 (1982).

8. Matsuzaki, M., Honkura, N., Ellis-Davies, G. C. R. & Kasai, H. Structural basis of long-term potentiation in single dendritic spines. *Nature* **429**, 761–6 (2004).

9. Tønnesen, J., Katona, G., Rózsa, B. & Nägerl, U. V. Spine neck plasticity regulates compartmentalization of synapses. *Nat. Neurosci.* **17**, 678–85 (2014).

10. Padmanabhan, P., Kneynsberg, A. & Götz, J. Super-resolution microscopy: a closer look at synaptic dysfunction in Alzheimer disease. *Nat. Rev. Neurosci.* **22**, 723–740 (2021).

11. Braitenberg, V. & Schüz, A. Cortex: statistics and geometry of neuronal connectivity. *Cortex Stat. Geom. Neuronal Connect.* (1998) doi:10.1007/978-3-662-03733-1.

12. Khanal, P. & Hotulainen, P. Dendritic spine initiation in brain development, learning and diseases and impact of BAR-domain proteins. *Cells* **10**, 2392 (2021).

13. Peters, A. & Kaiserman-Abramof, I. R. The small pyramidal neuron of the rat cerebral cortex. The perikaryon, dendrites and spines. *Am. J. Anat.* **127**, 321–55 (1970).

14. Melander, J. B. *et al.* Distinct in vivo dynamics of excitatory synapses onto cortical pyramidal neurons and parvalbumin-positive interneurons. *Cell Rep.* **37**, 109972 (2021).

15. Harnett, M. T., Makara, J. K., Spruston, N., Kath, W. L. & Magee, J. C. Synaptic amplification by dendritic spines enhances input cooperativity. *Nature* **491**, 599–602 (2012).

16. Kwon, T., Sakamoto, M., Peterka, D. S. & Yuste, R. Attenuation of synaptic potentials in dendritic spines. *Cell Rep.* **20**, 1100–1110 (2017).

17. Cornejo, V. H., Ofer, N. & Yuste, R. Voltage compartmentalization in dendritic spines in vivo. *Science (80-.).* (2021) doi:10.1126/science.abg0501.

18. Holler, S., Köstinger, G., Martin, K. A. C. C., Schuhknecht, G. F. P. P. & Stratford, K. J. Structure and function of a neocortical synapse. *Nature* **591**, 111–116 (2021).

19. Santuy, A., Rodriguez, J. R., DeFelipe, J. & Merchan-Perez, A. Volume electron microscopy of the distribution of synapses in the neuropil of the juvenile rat somatosensory cortex. *Brain Struct. Funct.* **223**, 77–90 (2018).

20. Berry, K. P. & Nedivi, E. Spine dynamics: are they all the same? *Neuron* **96**, 43–55 (2017).

21. Vardalaki, D., Chung, K. & Harnett, M.T. Filopodia are a structural substrate for silent synapses in adult neocortex. *Nature* **612**, 323–327 (2022).

22. LE, O. *et al.* Shifting patterns of polyribosome accumulation at synapses over the course of hippocampal long-term potentiation. *Hippocampus* **28**, 416–430 (2018).

23. Mancinelli, G. *et al.* Dendrite tapering actuates a self-organizing signaling circuit for stochastic filopodia initiation in neurons. *Proc. Natl. Acad. Sci. U. S. A.* **118**, e2106921118 (2021).

24. Ebrahimi, S. & Okabe S. Structural dynamics of dendritic spines: Molecular composition, geometry and functional regulation, *Biochim Biophys Acta - Biomembranes* **1838**(10), 2391–2398 (2014).

25. Dorkenwald, S. *et al.* Binary and analog variation of synapses between cortical pyramidal neurons. *Elife* **11**, e76120 (2022).

26. Spano, G. M. *et al.* Sleep deprivation by exposure to novel objects increases synapse density and axon–spine interface in the hippocampal CA1 region of adolescent mice. *J. Neurosci.* **39**, 6613–6625 (2019).

27. L, G., A, R. & M, H. Active dendrites enable strong but sparse inputs to determine orientation selectivity. *Proc. Natl. Acad. Sci. U. S. A.* **118**, e2017339118 (2021).

28. Campagnola, L. *et al.* Local connectivity and synaptic dynamics in mouse and human neocortex. *Science (80-.).* **375** (2022).

29. Loewenstein, Y., Yanover, U. & Rumpel, S. Predicting the dynamics of network connectivity in the neocortex. *J. Neurosci.* **35**, 12535–12544 (2015).

30. de Vivo, L. *et al.* Ultrastructural evidence for synaptic scaling across the wake/sleep cycle. *Science* **355**, 507–510 (2017).

31. Pfeiffer, T. *et al.* Chronic 2P-STED imaging reveals high turnover of dendritic spines in the hippocampus in vivo. *Elife* **7**, e34700 (2018).

32. Zito, K., Scheuss, V., Knott, G., Hill, T. & Svoboda, K. Rapid functional maturation of nascent dendritic spines. *Neuron* **61**, 247–258 (2009).

33. Noguchi, J., Matsuzaki, M., Ellis-Davies, G. C. R. & Kasai, H. Spine-neck geometry determines NMDA receptor-dependent Ca^{2+} signaling in dendrites. *Neuron* **46**, 609–622 (2005).

34. Aitchison, L. *et al.* Synaptic plasticity as Bayesian inference. *Nat. Neurosci.* **24**, 565–571 (2021).

35. Ólafsdóttir, H. F., Bush, D. & Barry, C. The role of hippocampal replay in memory and planning. *Curr. Biol.* **28**, R37–R50 (2018).

36. Krause, E. L. & Drugowitsch, J. A large majority of awake hippocampal sharp-wave ripples feature spatial trajectories with momentum. *Neuron* **110**, 722–733.e8 (2022).

37. Valero, M., Zutshi, I., Yoon, E. & Buzsáki, G. Probing subthreshold dynamics of hippocampal neurons by pulsed optogenetics. *Science (80-.).* **375**, 570–574 (2022).

38. Tripathy, S. J., Burton, S. D., Geramita, M., Gerkin, R. C. & Urban, N. N. Brain-wide analysis of electrophysiological diversity yields novel categorization of mammalian neuron types. *J. Neurophysiol.* **113**, 3474–3489 (2015).

39. Payeur, A., Guerguiev, J., Zenke, F., Richards, B. A. & Naud, R. Burst-dependent synaptic plasticity can coordinate learning in hierarchical circuits. *Nat. Neurosci.* **24**, 1010–1019 (2021).

40. P, R. *et al.* Projections from neocortex mediate top-down control of memory retrieval. *Nature* **526**, 653–659 (2015).

41. Goto, A. *et al.* Stepwise synaptic plasticity events drive the early phase of memory consolidation. *Science (80-.).* **374**, 857–863 (2021).

42. Vetere, G. *et al.* Spine growth in the anterior cingulate cortex is necessary for the consolidation of contextual fear memory. *Proc. Natl. Acad. Sci. U. S. A.* **108**, 8456–60 (2011).

43. O'Neill, J., Boccara, C. N., Stella, F., Schoenenberger, P. & Csicsvari, J. Superficial layers of the medial entorhinal cortex replay independently of the hippocampus. *Science* **355**, 184–188 (2017).

44. Payne, H. L., Lynch, G. F. & Aronov, D. Neural representations of space in the hippocampus of a food-caching bird. *Science (80-.).* **373**, 343–348 (2021).

45. Todorova, R. & Zugaro, M. Isolated cortical computations during delta waves support memory consolidation. *Science (80-.).* **366**, 377–381 (2019).

46. Bush, D., Freyja, H., Lafsdó, O., Barry, C. & Correspondence, N. B. Ripple band phase precession of place cell firing during replay. *Curr. Biol.* **32**, 64–73.e5 (2022).

47. Wang, D. V & Ikemoto, S. Coordinated interaction between hippocampal sharp-wave ripples and anterior cingulate unit activity. *J. Neurosci.* **36**, 10663–10672 (2016).

48. Rajasethupathy, P. *et al.* Projections from neocortex mediate top-down control of memory retrieval. *Nature* **526**, 653–659 (2015).

49. Zhang, H., Fell, J. & Axmacher, N. Electrophysiological mechanisms of human memory consolidation. *Nat. Commun.* **9**, 4103 (2018).

50. Wang, M., Foster, D. J. & Pfeiffer, B. E. Alternating sequences of future and past behavior encoded within hippocampal theta oscillations. *Science* **370**, 247–250 (2020).

51. Norman, Y. *et al.* Hippocampal sharp-wave ripples linked to visual episodic recollection in humans. *Science* **365**, eaax1030 (2019).

52. Sendhilnathan, N., Ipata, A. & Goldberg, M. E. Mid-lateral cerebellar complex spikes encode multiple independent reward-related signals during reinforcement learning. *Nat. Commun.* **12**, 6475 (2021).

53. Larry, N., Yarkoni, M., Lixenberg, A. & Joshua, M. Cerebellar climbing fibers encode expected reward size. *Elife*, **8**, e46870 (2019).

54. Csicsvari, J., Hirase, H., Mamiya, A. & Buzsáki, G. Ensemble patterns of hippocampal CA3-CA1 neurons during sharp wave-associated population events. *Neuron* **28**, 585–94 (2000).

55. Terada, S. *et al.* Adaptive stimulus selection for consolidation in the hippocampus. *Nature* **601**, 240–244 (2022).

56. Hodapp, A., Kaiser, M. E. & Thome, C., et al. Dendritic axon origin enables information gating by perisomatic inhibition in pyramidal neurons. *Science* **377** (6613), 1448–1452 (2022).

57. Bittner, K. C. *et al.* Conjunctive input processing drives feature selectivity in hippocampal CA1 neurons. *Nat. Neurosci.* **18**, 1133–1142 (2015).

58. Rueckemann, J. W., Sosa, M., Giocomo, L. M. & Buffalo, E. A. The grid code for ordered experience. *Nat. Rev. Neurosci.* **22**, 637–649 (2021).

59. Hebb, D. O. *The Organization of Behavior: a Neuropsychologial Theory.* (John Wiley and Sons, Inc., 1949).

60. Singer, A. C. & Frank, L. M. Rewarded outcomes enhance reactivation of experience in the hippocampus. *Neuron* **64**, 910–21 (2009).

61. Branco, T. & Staras, K. The probability of neurotransmitter release: variability and feedback control at single synapses. *Nat. Rev. Neurosci.* **10**, 373–383 (2009).

62. Fujisawa, S., Amarasingham, A., Harrison, M. T. & Buzsáki, G. Behavior-dependent short-term assembly dynamics in the medial prefrontal cortex. *Nat. Neurosci.* **11**, 823–33 (2008).

63. Mizuseki, K. & Buzsáki, G. Preconfigured, skewed distribution of firing rates in the hippocampus and entorhinal cortex. *Cell Rep.* **4**, 1010–21 (2013).

64. Aksoy-Aksel, A. & Manahan-Vaughan, D. The temporoammonic input to the hippocampal CA1 region displays distinctly different synaptic plasticity compared to the Schaffer collateral input in vivo: significance for synaptic information processing. *Front. Synaptic Neurosci.* **5**, 5 (2013).

65. Schapiro, A. C., Turk-Browne, N. B., Botvinick, M. M. & Norman, K. A. Complementary learning systems within the hippocampus: a neural network modelling approach to reconciling episodic memory with statistical learning. *Philos. Trans. R. Soc. B Biol. Sci.* **372**, 20160049 (2017).

66. Akers, K. G. *et al.* Hippocampal neurogenesis regulates forgetting during adulthood and infancy. *Science (80-.).* **344**, 598–602 (2014).

67. Franjic, D. *et al.* Transcriptomic taxonomy and neurogenic trajectories of adult human, macaque, and pig hippocampal and entorhinal cells. *Neuron* **110**, 452-469.e14 (2022).

68. Kohler, S. J., Williams, N. I., Stanton, G. B., Cameron, J. L. & Greenough, W. T. Maturation time of new granule cells in the dentate gyrus of adult macaque monkeys exceeds six months. *Proc. Natl. Acad. Sci. U. S. A.* **108**, 10326–10331 (2011).

69. Lee, J. Y. *et al.* Dopamine facilitates associative memory encoding in the entorhinal cortex. *Nature* **598**, 321–326 (2021).

70. Denoth-Lippuner, A. & Jessberger, S. Formation and integration of new neurons in the adult hippocampus. *Nat. Rev. Neurosci.* **22**, 223–236 (2021).

71. Tensaouti, Y., Stephanz, E. P., Yu, T.-S. & Kernie, S. G. ApoE regulates the development of adult newborn hippocampal neurons. *eneuro* **5**, ENEURO.0155-18.2018 (2018).

72. Yu, T.-S. *et al.* Astrocytic ApoE underlies maturation of hippocampal neurons and cognitive recovery after traumatic brain injury in mice. *Commun. Biol.* **4**, 1303 (2021).

73. Motta, A. *et al.* Dense connectomic reconstruction in layer 4 of the somatosensory cortex. *Science (80-.).* **366**, eaay3134 (2019).

Chapter 6

Sleeping and Dreaming

Abstract

Humans and many animals sleep every day. We are now closer to a coherent understanding of how sleeping and dreaming facilitates learning and refreshes the brain. Harmful wastes are cleaned in deep sleep. New neurons and new dendritic spines are shaped where they belong. The weights of synapses are adjusted as an animal sleep, and some synapses are permanently lost. The overall network is stable and evolving for decades. Computers do not have to use exactly the same way, but the principles can probably help.

Keywords

Glymphatic circulation, Cerebrospinal fluid (CSF), Non-rapid eye movement (NREM) sleep, Slow-wave sleep, Interneurons, Theta waves, acetylcholine, Rapid eye movement (REM) sleep, Memory replay, Daydreaming

6.1 Non-Rapid Eye Movement (NREM) Sleep—Flushing Waste Out of the Brain and Stock-Up

Sleeping, especially slow-wave sleep, is integral to hippocampal and hypothalamic neurogenesis[5, 6] (Fig. 4.13). Cerebrospinal

Neuroscience for Artificial Intelligence
Huijue Jia
Copyright © 2023 Jenny Stanford Publishing Pte. Ltd.
ISBN 978-981-4968-78-2 (Hardcover), 978-1-003-41098-0 (eBook)
www.jennystanford.com

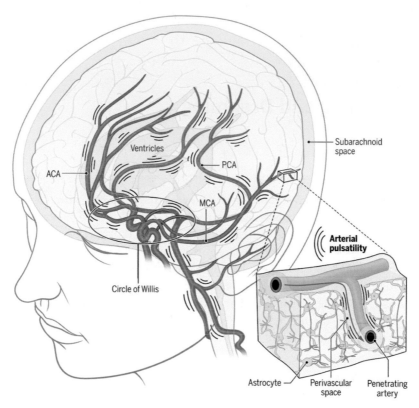

Figure 6.1 Arterial pulsatility propels fluid flow in the brain. The brain receives 20 to 25% of a person's cardiac output but constitutes only ~2% of total body weight. The large-caliber arteries of the circle of Willis are positioned in the cerebralspinal fluid (CSF)-containing basal cisterns below the ventral surface of the brain. Arterial pulsatility provides the motive force for CSF transit into the perivascular spaces surrounding the major arteries, whereas respiration and slow vasomotion contribute to sustaining its flow[1]. The anterior (ACA), middle (MCA), and posterior (PCA) arteries transport CSF to the penetrating arteries (inset), from which CSF is then driven into the neuropil via the still-contiguous perivascular spaces. Cardiovascular diseases associated with reduced cardiac output, such as left heart failure and atrial arrhythmias, reduce arterial wall pulsatility, resulting in diminished CSF flow. In addition, thickening of the arterial wall in small vessel disease, hypertension, and diabetes reduces arterial wall compliance and, hence, pulsatility. Each of these fundamentally cardiovascular disorders serves to attenuate glymphatic flow, providing a potential causal link between these vascular etiologies and Alzheimer's disease[2].

Credit: Fig. 4 of ref. 3.

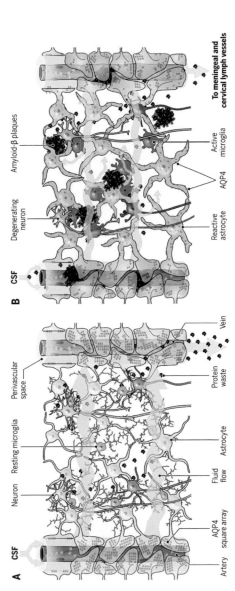

Figure 6.2 The brain glymphatic system is a highly organized fluid transport system. (A) Vascular endfeet of astrocytes create the perivascular spaces through which CSF enters the brain and pervades its interstitium. CSF enters these perivascular spaces from the subarachnoid space and is propelled by arterial pulsatility deep into the brain, from where CSF enters the neuropil, facilitated by the dense astrocytic expression of the water channel aquaporin-4 (AQP4), which is arrayed in nanoclusters within the endfeet. CSF mixes with fluid in the extracellular space and leaves the brain via the perivenous spaces, as well as along cranial and spinal nerves. Interstitial solutes, including protein waste, are then carried through the glymphatic system and exported from the CNS via meningeal and cervical lymphatic vessels. (B) Amyloid-β plaque formation is associated with an inflammatory response, including reactive micro- and astrogliosis with dispersal of AQP4 nanoclusters. Age-related decline in CSF production, decrease in perivascular AQP4 polarization, gliosis, and plaque formation all impede directional glymphatic flow and thereby impair waste clearance. Notably, vascular amyloidosis might be initiated by several mechanisms. Amyloid-β might be taken up from the CSF by vascular smooth muscle cells expressing the low-density lipoprotein receptor-related protein 1 (LRP1)[4]. Alternatively, amyloid deposition might be initiated by the backflow of extracellular fluid containing amyloid-β into the periarterial space from the neuropil, rather than proceeding to the perivenous spaces, because of an increase in hydrostatic pressure on the venous side or an inflammation-associated loss of AQP4 localization to astrocytic endfeet.

Credit: Fig. 1 of ref. 3.

fluid (CSF), which comes from plasma, flows in to exchange with interstitial fluid (Figs. 6.1 and 6.2), and clears interstitial molecules including the β-amyloid (Aβ) protein that is known to accumulate in Alzheimer's disease patients[7]. Molecules from periphery tissues, such as serotonin and other molecules from the gut (Section 6.4), may get into the central nervous system during this extensive exchange. Sleep-deprived adolescent mice had more and larger dendritic spines in hippocampal CA1 neurons, while the presynaptic boutons contained a smaller proportion of vesicles available for release[8]; Spontaneous awake mice, on the other hand, had a smaller total number of vesicles, while the proportion of available vesicles were not different from mice that had been sleeping[8].

Within each neuron that go across cortical layers (Fig. 1.9), the useful materials might redistribute (concomitant with a decreased size of many postsynaptic dendritic spines) and get ready for the responses required the next day. This also means that neurons that receive more synaptic input could accumulate more resources over time, a feedforward mechanism that helps it stand out from its neighbors (reminiscent of Fig. 3.5).

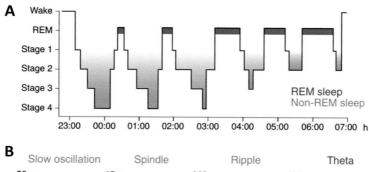

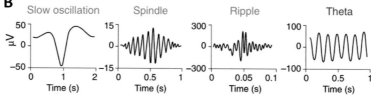

Figure 6.3 Sleep architecture and sleep oscillations. (A) Human nocturnal sleep profile. Compare with Chapter 3, Fig. 3.7. (B) Neocortical slow oscillations, thalamocortical spindles and hippocampal ripples are electrophysiological signatures of non-REM sleep. Theta oscillations are prominent during REM sleep, particularly in the rodent (rats or mice) hippocampus.

Credit: From Box 3 of ref. 6.

6.2 The Alternating and Progressing Phases of Sleep

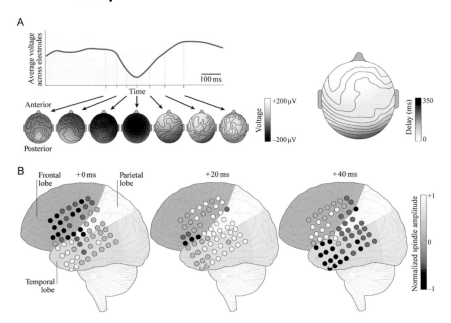

Figure 6.4 Macroscopic waves during human sleep. (A) The slow oscillation of deep non-rapid-eye-movement (non-REM) sleep has been reported to be a traveling wave that moves globally from anterior to posterior regions[10]. The upper left panel represents the time course of one slow oscillation averaged across electroencephalogram (EEG) channels. The lower left panels illustrate the evolution of voltages across the scalp at different times during the slow wave. At the start of the slow oscillation, negative EEG potentials begin in anterior sections of the scalp, and as the oscillation progresses, these travel to posterior regions. The negative peak of the slow oscillation is delayed in posterior regions by hundreds of milliseconds, reflecting the propagation time of the traveling wave (right panel). (B) Sleep "spindles" are 11–15 Hz oscillations that occur during stage 2 non-REM sleep and have long been known to be important for learning and memory. These spindles systematically travel as global rotating waves from the temporal to the parietal to the frontal cortex (and are therefore called TPF waves) in intracranial electrocorticography (ECoG) recordings[11]. The panels illustrate the normalized spindle amplitude at each ECoG electrode (small circles) at successive points in the spindle cycle. From the start of an oscillation cycle (+0 ms), spindle amplitudes peak successively in the temporal (+0 ms), parietal (+20 ms) and frontal (+40 ms) lobes. The amplitudes recorded for each channel are normalized to their maximum amplitude.

Credit: Fig. 1 of ref. 12. Part (A) was republished with permission from ref. 10. Part (B) was adapted by ref. 12 with permission from ref. 11.

Sleeping drive increases toward the end of the day, carrying plenty of waste and low on the activating neurotransmitter acetylcholine and presynaptic vesicles. Slow-wave sleep, i.e., non-rapid eye movement sleep (NREM), dominates early at night (Fig. 6.3). An intriguing possibility is that the waves of activity impact blood flow to then draw in CSF flow for waste clearance[9], especially at the beginning of NREM sleep (Figs. 6.3, 6.4, 6.5 and 3.9).

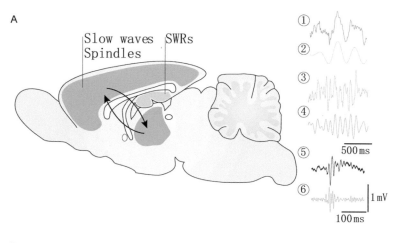

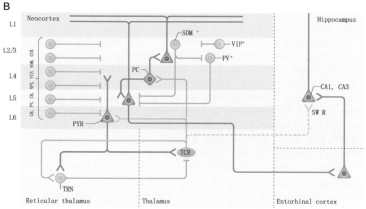

Figure 6.5 Circuit mechanisms of NREM sleep-specific oscillations. During non-rapid eye movement (NREM) sleep, slow waves predominantly propagate across the neocortex in an anterior-to-posterior direction in mice and humans, and decrease in amplitude over time[15] (Fig. 6.3). (A) Representative 1 s samples of NREM events recorded in mice: (1) raw cortical electroencephalography

(Continued)

Figure 6.5 (*Continued*)

(EEG); (2) slow waves (1.0–4.5 Hz); (3) a single sleep spindle in the raw cortical EEG; (4) a band-filtered spindle trace (11–15 Hz); (5) hippocampal raw local field potentials; (6) sharp-wave ripples (SWRs; 100–250 Hz). Cortical EEG recordings during NREM sleep are characterized by slow waves, which are synchronized with sleep spindles in thalamocortical circuits and SWRs in the hippocampus and are important for sleep integrity and sleep-dependent memory formation. (B) The circuit mechanisms of oscillations occurring during NREM sleep. Slow-wave oscillations reflect recurrent excitatory inputs between the thalamus and neocortex, which are shaped by inhibitory interneurons in layers 2/3 and 5 of the neocortex[16]. Excitatory inputs from both the neocortex and thalamus to inhibitory thalamic reticular nucleus (TRN) neurons drive a hyperpolarizing input to thalamocortical relay (TCR) cells, which causes rebound burst firing that sends a volley of excitatory activity to the neocortex, giving rise to sleep spindles[17]. At the same time, collateral inputs to hippocampal area CA1/CA3 as well as inputs from the entorhinal cortex drive SWRs[18], which co-occur with spindles. AN, anterior nuclei; CB, calbindin; CCK, cholecystokinin; CM, centromedial nucleus; CR, calretinin; IL, internal (medullary) lamina; LP, lateral posterior nucleus; MD, mediodorsal nucleus; NPY, neuropeptide Y; PC, principal neuron; PFC, prefrontal cortex; PV, parvalbumin; PYR, pyramidal neuron; SOM, somatostatin; VA, ventral anterior nucleus; VIP, vasoactive intestinal peptide; VL, ventrolateral nucleus; VPL/VPM, ventral posterior lateral nucleus/ventral posterior medial nucleus. PV, CB, CR are all calcium-binding proteins. For the classification of interneurons, see also Fig. 6.6.

Credit: Fig. 2b,c of ref. 19.

The inflow of CSF and the resulting osmolarity change likely allow the neurons (both old and new) to enlarge, which would release at synapses more easily[13], and seen as sharp-wave ripples (SWRs, Section 5.4). Learning before the NREM sleep increases coupling between slow oscillation, sleep spindle (Fig. 6.4), and SWRs (Figs. 6.3 and 6.5). Neocortical sleep spindles were observed to entrain hippocampal spindles, and precede SWRs in the hippocampus (the neocortical spindle started 200–250 ms before and the hippocampal spindle 50–100 ms before the SWRs)[14], consistent with memory consolidation with hashing (Fig. 4.11).

Occipital slow waves occur with a similar density across all sleep stages[20]. A transient increase in dopamine in the basal amygdala during NREM sleep initiates REM sleep[21]. Proportion of REM sleep increases towards the end of the night and the probability of waking up increases (Fig. 6.3), e.g., waking up when trying to mobilize muscles (the supplementary motor area or the parietal cortex[22]) to do something.

6.3   Interneurons—Global or Local Patterning with Brain-Wide Oscillations

The information content (and entropy) within a population is reduced during synchronous activity[23]. Unexciting images (reminiscent of Fig. 5.10) increased synchronized γ oscillations in the monkeys' visual cortex[24]. The brain-wide oscillations likely pattern and reset the network during NREM sleep. But synapses with large spine heads will likely survive all the remodeling (Chapter 5, Fig. 5.7), while new synapses are selectively nurtured.

Interneurons are a minority group compared to the pyramidal neurons (~89% of the synapses are excitatory, Fig. 5.4)[25], but the brakes they exert are key to synchronous activities (Figs. 1.9 and Fig. 6.5, 6.6)[26, 27]. According to modeling using a recurrent neural network, the inhibitory interneurons are likely more important for learning more general features, while the excitatory pyramidal neurons encode objects[28]. But a study in the primary visual cortex found connections between inhibitory neurons and between excitatory neurons within the same layer to show a larger spatial distance than connections between an inhibitory neuron and an excitatory neuron in either direction within the same layer[29].

Schizophrenia patients showed decreased expression of the calcium-binding protein parvalbumin (PV) in PV-expressing interneurons[30]. Acute stress activated somatostatin-expressing interneurons in the prefrontal cortex, in favor of amygdala-driven feedforward inhibition via the interneurons[31].

Traditional classification of interneurons based on shape (morphology) and connections are being reconciled with classification based on recent technologies such as single-cell transcriptomes[32, 33] (Fig. 6.6). Martinotti cells express somatostatin (abbreviated as SOM or SST), classic (large) basket cells (Fig. 6.7) and chandelier cells express PV, small basket cells and other cells express vasoactive intestinal peptide (VIP, Figs. 6.5 and 6.6)[32–34].

These expressing the calcium-binding protein parvalbumin are fast-spiking[34, 35] (we will also see them in Chapter 7) and their action potentials can go backward from the axons to their dendrites, i.e., backpropagation[36, 37]. Through a PV-expressing interneuron (or >100 ms with a Martinotti cell), for example, a pyramidal neuron in layer II can inhibit another pyramidal neuron in layer II or III in 3–6 ms[29].

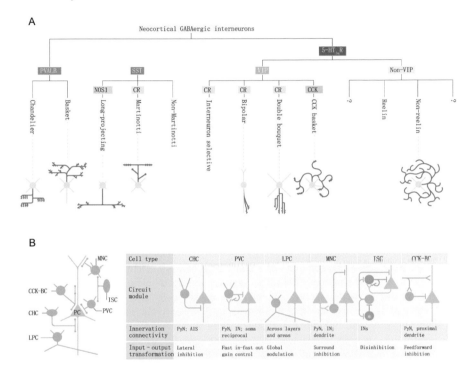

Figure 6.6 A work draft taxonomy of transcriptomic neuron types of the cortical GABAergic system and their interaction with pyramidal neurons. (A) Major GABAergic subclasses and cell types recognized by classic anatomical, physiological, molecular and developmental studies. Schematics of the characteristic morphology of these cell types are also shown. Light gray lines represent dendrites; dark gray lines represent axons. 5-HT$_{3A}$R, serotonin 3A receptor; CCK, cholecystokinin; CR, calretinin; NOS1, nitric oxide 1; PVALB, parvalbumin; SST, somatostatin; VIP, vasoactive intestinal peptide. (B) Six cardinal types — long projection cells (LPCs), chandelier cells (CHCs), PV-basket cells (PVCs), cholecystokinin expressing basket cells (CCK-BCs), interneuron selective cells (ISCs) and Martinotti cells (MNCs) — are distinguished by their characteristic innervation of cellular and subcellular targets, i.e., dendrites, cell body, axons. Where data are available, these cardinal types manifest distinct input–output connectivity patterns and further display distinct intrinsic, synaptic and network properties indicative of mediating specific forms of input–output transformation.

Credit: Fig. 1a, Fig. 2a of ref. 32.

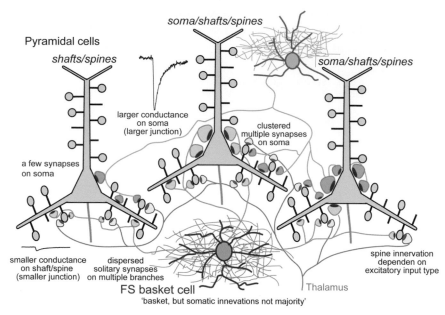

Figure 6.7 Schematic summary of a study on inhibitory synapses from basket cells to crossed-corticostriatal (CCS) "slender untufted" pyramidal cells in L5 of the rat cortex[35]. Among ten pairs of basket cells and pyramidal cells analyzed (each pair showing 5–14 synapses), three pairs showed synapses on the soma and dendrites, and seven pairs showed synapses on dendrites only. The junctional area was 0.194–0.350 μm^2 for synapses on the cell body (soma) of pyramidal cells, 0.102 μm^2 for synapses on dendritic shafts and 0.042–0.056 μm^2 for synapses on dendritic spine heads. Axonal bouton volume was linearly correlated with synaptic junction area, consistent with Chapter 5.

Credit: Fig. 7 of ref. 35.

Dopamine receptor-expressing neurons from the mediodorsal nucleus of the thalamus preferentially innervate VIP-expressing interneurons in the prefrontal cortex, which enable decisions based on faint evidence, i.e., sparse task-relevant cues; Thalamic neurons expressing glutamate receptors (kainite receptors), on the other hand, innervate PV-expressing interneurons in the prefrontal cortex to facilitate decisions under noisy/conflicting situations[38, 39].

Martinotti cells in cortical layers II-IV target distal dendrites in layer I and also more proximal sites (Figs. 6.6B and 6.5); Martinotti cells often have an elaborate dendritic tree and the presynaptic boutons are also spiny[32, 34] (Fig. 6.6A). The presynaptic vesicles may contain both glutamate and GABA (γ-aminobutyrate)[40], which could act on pyramidal neurons and other interneurons at the same time. GABA could be excitatory for pyramidal neurons during development, before switching to become an inhibitory signal[41].

Basket cells target cell bodies or proximal dendrites (Fig. 6.7), and chandelier cells target axon initial segments of pyramidal neurons[32, 34] (and function in a different phase, Section 7.8). Basket cells can also inhibit neighboring basket cells, double bouquet cells, etc.[34] These different types of interneurons could lead to differences in computation, memory and resource allocation in the pyramidal neurons. Fast-spiking basket cells (PV-expressing) in layer V of the rat cortex, for example, appeared to have smaller contact areas on proximal dendritic spines and shafts of pyramidal neurons compared to their contact areas on the cell body (Fig. 6.7), corresponding to a smaller electric charge[35]. Such a weaker effect on dendritic spines would make these inhibitory synapses disadvantaged during cellular computation (Chapter 5), even before their greater distance from the cell body was considered, while being more specific and could counter an excitatory input from the thalamus on the same dendritic spine (Fig. 6.7)[35, 41].

Pyramidal neurons form about six synapses onto a classic (large) basket cell, with around 60% on "basal" dendrites, 30% on the main dendrite and 10% on the soma [27]. In circuits such as feedforward or feedback inhibition, information kept in the dendritic spines, shafts, cell body and axons of the interneurons is also updated with activity[32, 34, 42] (Fig. 6.6).

We mentioned in Chapter 3 that older people do not sleep as deeply, i.e., shorter NREM. Healthy adults showed more NREM sleep during the recovery sleep after total sleep deprivation, which accumulated tau and amyloid-β already[9]. A bed rocking at 0.25 Hz can increase the duration of NREM3 sleep and increase fast spindles in synchrony with slow oscillations (Fig. 6.5A), facilitating declarative memory[43].

Some neurons are slower than others[44]. After attributing SWRs to excitatory synapses in Chapter 5, it follows that cortical γ oscillations might be due to interneurons that express parvalbumin[26, 27] (basket cells or chandelier cells[32, 34], Figs. 6.5 and 6.6), which are the most common type and the type with the highest fraction of electrical connections (gap junctions) versus chemical connections[29]. A wilder guess would be that the slower waves in NREM sleep represent synaptic activity in the presence of oscillating fluids (Fig. 6.1; see the REM Section 6.5 for θ oscillations due to postinhibition "rebound" activation), or other noncortical neurons that are even slower[44]. The rate of breathing falls in the range of slow oscillations (0.1–1 Hz, Fig. 6.1). Sleep spindles (~11–15 Hz, involving the thalamus, Figs. 6.4 and 6.5)[19] have been implicated in declarative memory formation when learning a second language[45, 46] (Section 8.7.2), which would be forming the hashes (Fig. 4.11).

Compared to dreams, hallucinations in slow-wave sleep are usually less visual, more static and without narratives, perhaps because the interplay between entorhinal cortex and hippocampus is different from both wakefulness and REM sleep (more on space and time in Chapter 7), so that the SWRs can more repetitively work on the weight of individual synapses and remember things (Chapter 5).

6.4 Evolutionarily Ancient Circuits Tapping into Our Dreams?

Relatively independent functions likely modulate the sleeping and dreaming cycle in the brain. In mice treated with anesthesia regimens, heart rate negatively correlated with CSF influx, while systolic blood pressure might have a positive correlation[47] (Fig. 6.1). Respiration entrains higher-frequency oscillations in the olfactory bulb, piriform cortex (PC), amygdala, and hippocampus[48]. Awake volunteers recognized fearful faces faster when inhaling compared to exhaling (like a mouse sniffs danger[49]), a phenomenon that was only observed when breathing through the nose, and not through the mouth (Fig. 6.8).

The body temperature drops by ~1°C during sleep, more in men than in women[50, 51], and the glymphatic flow may help cool off the brain (Fig. 6.1). The gut microbiome in women likely produces more acetate than in men[52]. Butyrate peaks after sleep[53, 54] and could increase NREM sleep[55], while acetate possibly peaks closer to wake-up time (Fig. 6.3). Hormonal changes with age, before menses and during pregnancy, i.e., lack of testosterone, high progesterone and follicle-stimulating hormone (FSH), contribute to differences in body temperature[56, 57], which could influence the microbes that live in our body[58–62] and possibly their metabolites such as acetate. In flying bats, the body temperature reaches 40°C. It was shown in lizards that a higher body temperature makes δ/β oscillations faster[63]; In lizards they normally take almost 2 min[63] and are about two orders of magnitude slower than our δ/β oscillations (Fig. 6.5). Neuronal functions such as vision are also impacted by body temperature[64].

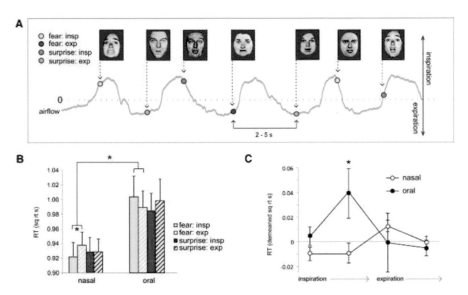

Figure 6.8 Respiratory phase modulates fear-related response times. A, Emotion discrimination task. Subjects viewed faces expressing either fear or surprise, and indicated which emotion was perceived. Interstimulus interval, 2–5 s. Colored dots indicate where in the breathing cycle stimuli were encountered. B, Fearful faces were detected more quickly during nasal inspiration vs expiration, but not during oral breathing. C, Emotion RT data, binned across four phases of breathing, revealed a significant two-way interaction between breathing time bin (4 levels) and breathing route (nasal/oral) for fearful faces, with maximal RT differences during nasal fear trials occurring between the onset of inspiration and the onset of expiration. *$p < 0.05$ in all panels. Error bars represent the SEM.

Credit: Fig. 8 of ref. 48.

In addition to acetate and choline, neurotransmitters such as dopamine, GABA (γ-aminobutyrate), serotonin and other metabolites of tryptophan are also produced or metabolized by the microbiome, which likely contribute to diseases such as schizophrenia[65-67].

6.5   Rapid Eye Movement (REM) Sleep

Octopus show <1 min unresponsive periods with active twitching of eyes, muscles and sucker pads, between their ~7 min quiet sleep periods which have paler skin color and closed pupils[68]. Another order of distantly related mollusks, cuttlefish, also have REM-like periods of rapid eye movements, color change and arms twitching[69]. As we note in Chapter 1, the human brain consumes a lot of energy. Although actual (but not simulated) vision and locomotion are shut off, the brain's metabolic rate is no lower during REM sleep than during waking hours[70], and 15% less during NREM sleep[3]. This is either due to the brain-wide extent of the remodeling activity, or due to fluids remaining from the previous NREM that renders electric activities "leaky." NREM sleep also contain short periods of dreaming, when the slow waves are weak in some region[70, 71].

We mentioned in Chapter 2 that the thalamus (and amygdala, hippocampus) has axons going to dendrites in Layer I of the neocortex and sends information such as body movements. The thalamus is active during both NREM and REM sleep (Figs. 6.5 and 6.9), despite the person being immobile and not responding to external stimuli.

Dreaming associates with visuo-spatial skills in children[70] (Chapter 2, Chapter 7). Despite a high proportion of REM sleep in infants (e.g., for effective development of motor functions[73-75]), vivid dreaming develops later. The parietal lobes did not become fully myelinated until age 7. Imagined navigation in healthy adults show a grid-like pattern (Chapter 7) in brain activity, just like navigation in virtual reality[76].

Interregional path in the cortex appeared lengthened in patients of Alzheimer's disease[77]. We cannot directly see in humans whether or not REM and NREM sleep eliminate convoluted neuronal wiring for sparse but reliable local and multi-modal connections. Traveling waves of spontaneous activity are known to pattern the development of the visual system[78].

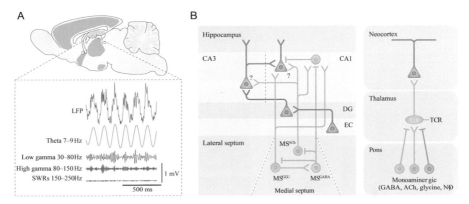

Figure 6.9 Circuit mechanisms of REM sleep-specific oscillations. (A) Representative 1s traces of mouse hippocampal local field potentials (LFPs) recorded during rapid eye movement (REM) sleep: raw LFPs, filtered theta activity (7–9 Hz), low-gamma activity (30–80 Hz), high-gamma activity (80–150 Hz) and sharp-wave ripples (SWRs; 150–250 Hz). REM sleep is characterized by the predominance of theta oscillations and the appearance of gamma oscillations in both LFP and cortical electroencephalography signals. Note the coherence (phase-locking) of gamma power to the ongoing theta oscillation during REM sleep. In addition to theta and gamma oscillations, the brainstem generates ponto-geniculo-occipital waves in humans (not shown), which have been suggested to contribute to visual perception during dreaming as well as memory extinction. (B) Circuit mechanisms underlying REM sleep oscillations in hippocampus and cortical networks. The medial septum (MS) contains genetically distinct cells producing acetylcholine (ACh), glutamate and GABA whose ongoing activity is essential for the generation of theta oscillations in areas CA1 and CA3 of the hippocampus (left panel), together with other inputs from the entorhinal cortex (EC), dentate gyrus (DG) and subcortical structures. GABAergic cells (and to a lesser extent, glutamatergic neurons) in the MS also contribute to theta activity in the hippocampus; however, the precise circuit mechanism and the role of ACh remain unclear. Ponto-geniculo-occipital waves during REM sleep result from bursts of monoaminergic inputs in the brainstem to the lateral geniculate nucleus of the thalamus (right panel), which in turn relays the volley to the visual cortex. NO, nitric oxide; TCR, thalamocortical relay neuron.

Credit: Fig. 3 of ref. 19.

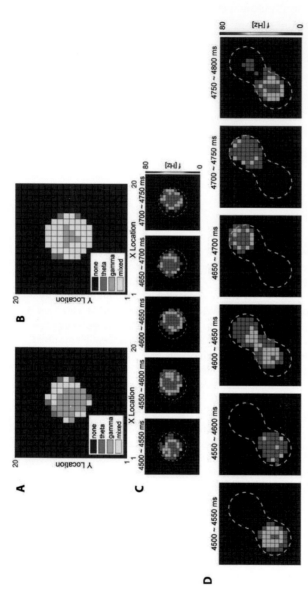

Figure 6.10 Simulating acetylcholine-driven gamma (30–100 Hz) and theta (5–12 Hz) band cell firing with one or two hotspots. (A,B,C) Example single peak *gKs* mapping projected on the 20×20 Excitatory (E) cell lattice. The neurons are colored according to firing frequency in the E-I (Inhibitory) network simulation[72]. There are 400 E neurons and 100 I neurons. Green, primarily exhibiting gamma band firing (Fig. 6.7); Light blue, theta; Yellow, mixed rhythms; Dark blue, none. (A,B) Dominant rhythmic activity of individual E cells plotted at cell position on the E cell lattice for *gKs* hotspot radius of *r* = 5.5 and *r* = 6.1, respectively. (C) Snapshots of E cell firing activities from 4500–4750 ms during the simulation shown in B (*gKs* hotspot radius *r* = 6.1). (D) Example of double peaked spatial mapping of *gKs* values for corresponding neurons on the 20×20 E cell lattice for hotspot radius *r* = 6.1 and distance between hotspot centers *d* = 8. Snapshots of E cell firing rates from 4500–4800 ms during the simulation. Cells inside the orange contour lines have values less than 0.6 mS/cm².

Credit: Fig. 3C,D,E and Fig. 4I of ref. 72.

Driven by the brainstem (Fig. 6.9), REM sleep shakes up inputs during the day that have led to tentative dendritic growth and reduces the total number of synapses (and to proceed through time, Section 7.3). New neurons and new dendritic spines that grow during NREM sleep might also test run with existing circuits during REM sleep. With the cells relatively in place and straightened by the flow, the dendrites might slip on some contact surfaces during the alternating phases of NREM and REM sleep, without losing contact entirely (like we holding a hot potato, or a kelp holding onto a rock). Such a process might also normalize the network across time[79], e.g., starting from a high reference point (e.g., a large spine[80]) and adjust the other ones accordingly.

New neurons added to the hippocampus for learning a particular experience do get activated in REM sleep; and if silenced, the neck of their dendritic spines became longer and the experience was not effectively remembered[81] (consistent with Chapter 5).

As we see in Fig. 5.7, synapses with larger dendritic spine heads are more likely to survive after a few days[80]. So this is not determined by the neck of a postsynaptic spine (which can insulate the signal within a spine, Figs. 5.2 and 5.3)[82, 83], but the volume of current once activated, which corresponds to a larger certainty of synaptic transmission (Section 5.2.2). Even though the neurons likely accumulate resources with synaptic input, if the amounts of structural molecules in each neuron is relatively constant, to enlarge a dendritic spine would still mean to make some other spines smaller or demolished[84]. New spines likely compete with each other for existing presynaptic boutons, before a mature synapse is formed[83, 85]. With new spines, more than one spine could be in contact with their presynaptic partner, making them an interesting candidate player in the content of dreams. The package of neurotransmitters received might also get passed on within a neuron and sequentially between neurons, to influence regions deeper down, if there is still some left.

Bursts of neurotransmitters from the brainstem is relayed by the thalamus to the visual cortex (Fig. 6.9). Acetylcholine is the major activating neural transmitter during REM sleep, whereas the levels of other neural transmitters (norepinephrine, serotonin, histamine, cretin) are greatly reduced[70].

Localized release of acetylcholine (synthesized from acetyl-Coenzyme A and choline, related to Section 6.4) is known to promote θ-γ coupling, and generates hotspots of certain radii in the cortex[72]. According to simulation with an excitation-inhibition network, the center showed γ oscillation, which weakened to θ oscillation at the edge[72] (Figs. 6.10 and 6.9). Inhibition of pyramidal neurons by GABAergic interneurons (Figs. 6.6 and 6.7) is also responsible for θ oscillations in the hippocampus[86]. θ band frequency increased when the excitatory external current input was greater, and when the inhibitory synapses had a faster decay time constant[72]. When the acetylcholine excitation was smaller than the hotspot (e.g., Fig. 4.5), the activation would still play out in the entire hotspot[72]. Two or more nearby hotspots can alternate in activity (Fig. 6.10D). Visual navigation, for example, takes place with electric signals in γ and SWR frequency[87–90] (Fig. 6.9; Hippocampal SWRs are slower in humans likely due to the larger synapses than in rats). Besides dreaming, such acetylcholine-driven spontaneity may also be involved in diseases such as schizophrenia[66, 91] and posttraumatic stress disorder (PTSD).

Hippocampal θ waves are active during REM sleep as during wakefulness (Figs. 6.11 and 6.12), which is one of the candidate types of waves for encoding time series (Fig. 6.13, Chapter 7), and are lacking in the storyless hallucinations typical of NREM sleep. Cortical interneurons that express parvalbumin are involved in the generation of γ oscillations[26, 27] (Figs. 6.9 and 6.10, Section 6.3), which are phase-locked with hippocampal waves[92, 99]. The oscillations are also a sampling strategy[100] (Section 4.4). After waste clean-up and neuronal/dendritic growth during NREM sleep, hippocampal θ waves or more spontaneously from the cortex itself, likely test play neocortical information in REM (Figs. 6.10 and 6.11, feedback hashing model in Section 4.4). The character in one's dream can shift from one person to another (perhaps similar to Fig. 6.10); some sensory information (e.g., pressure on limbs, odor) could emerge as part of the story, according to established links[101].

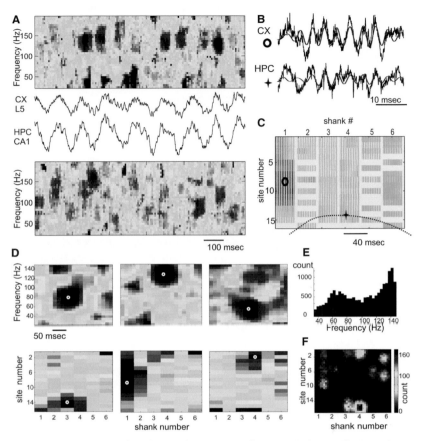

Figure 6.11 Temporal and spatial structure of neocortical γ oscillations that were preceded by hippocampal θ waves[92]. See also[93, 94]. (A) A short epoch of neocortical (CX L5) and hippocampal (HPC CA1) LFPs and their associated "whitened" spectrograms. (B) Gamma "burst" (red, band-pass, 100–200 Hz, signal) from sites shown in (C). (C) Color-coded spatial profile of band-pass-filtered segment in (B) at all recording sites (anatomical layout as in Fig. 3A). Each column, separated by gray vertical stripes, corresponds to an electrode shank with 16 recording sites each. Malfunctioning sites are gray. (D) Examples of isolated gamma bursts in hippocampus (left) and neocortex (middle, right). Each burst is characterized by a local maximum (white circles) of LFP spectral power (color) in both time-frequency (top) and anatomical space (bottom). (E) Distribution of frequencies of individual local maxima. Note two modes, slow and fast gamma. (F) Probability density of the spatial locations of local maxima of gamma power for the entire session. Note spatially segregated clusters.

Credit: Fig. 4 of ref. 92.

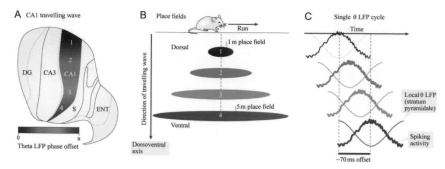

Figure 6.12 Patterning of place field firing by hippocampal traveling waves. (A) The 6–12 Hz hippocampal theta oscillation, which appears during active behavioral exploration and sleep, is a wave traveling from dorsal to ventral hippocampal CA1[95, 96]. This flattened map view of the rodent hippocampus shows the positions of the dentate gyrus (DG), CA3, CA1, subiculum (S) and entorhinal cortex (ENT). The phase offset of the theta oscillation is plotted in color along the dorsoventral axis of CA1 (color axis, bottom left). (B) As the rodent runs on a linear track, pyramidal neurons in CA1 fire. These neurons exhibit place fields that become progressively larger as their location in CA1 changes from dorsal to ventral regions[97, 98]. The schematic illustrates a typical example of the place fields of neurons that fire when the rodent is at a particular spatial location during the run and is based on data in ref. 95. The place fields are arranged according to their location in CA1 and with their color illustrating the phase offset of the theta oscillation at that location on the dorsoventral axis. (C) Because of the phase offset of the theta local field potential (LFP; as recorded in stratum pyramidale[96]; shown as light transparent lines) owing to the traveling wave (represented by the distance between the dotted lines) and the modulation of neuronal firing by theta (solid lines), neurons with CA1 place fields centered at the rodent's position fire in a temporal sweep from those with the smallest scales of spatial selectivity to those with the largest scales of spatial selectivity.

Credit: Fig. 5 of ref. 12. Part (A) was reproduced from ref. 95. Part (C) was adapted by ref. 12 with permission from ref. 96.

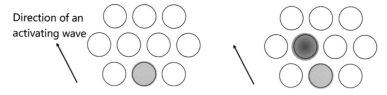

Figure 6.13 Waves of activity might convert the temporal order into spatial order in information storage or retrieval. Also relevant for searching and hashing (Fig. 4.14), and possibly language functions (Chapter 8).

Credit: Huijue Jia.

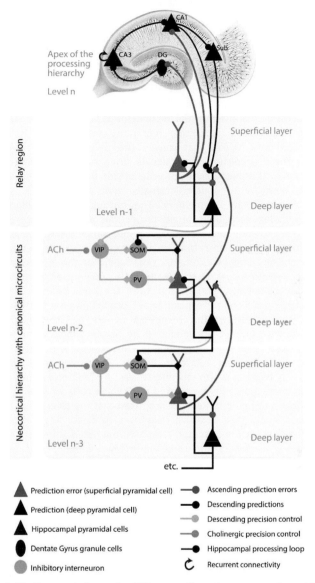

Figure 6.14 Proposal by ref. 102 regarding the neuronal architecture underlying inhibitory and facilitatory hippocampal-neocortical interaction. Within the neocortical hierarchy, message passing is orchestrated by a canonical microcircuit that includes both excitatory (red and black) and inhibitory (beige) cells. In the superficial layers of each cortical level, superficial pyramidal cells (red) compare the activity of representational units (black) with top-down predictions relayed via SOM+ inhibitory interneurons

(*Continued*)

Figure 6.14 (*Continued*)

(SOM). These interneurons are targeted by descending prediction signals that originate in deep pyramidal cells (black) from the level above. The mismatch between representations and descending predictions (black lines) constitutes a prediction error. This prediction error signal (red lines) is passed back up the cortical hierarchy and is received by prediction units (black) that drive responses in higher representational units, or, at the apex of the processing hierarchy, in the hippocampus. Therefore, as information moves up the cortical processing hierarchy, sensory input is replaced by prediction error signals that convey the only information yet to be explained. These prediction error signals drive representations in higher levels of the cortical hierarchy to provide better predictions, but also drive associative plasticity to update internally generated predictions that in the hippocampus draw on memory. The output from the hippocampus targets neocortex via glutamatergic projections to deep pyramidal cells (black, e.g., in the entorhinal cortex), or via long-range GABAergic projections to superficial cells (e.g., in retrosplenial cortex; not shown here). Using a predictive coding framework, ref. 102 proposed that the hippocampus uses a unitary code with a dual aspect function. This dual aspect function can be characterized as follows: During prediction, the hippocampus can provide multi-sensory predictions to "explain away" prediction errors at lower levels of the cortical hierarchy. This manifests as an inhibitory hippocampal-neocortical interaction – here mediated by SOM+ inhibitory interneurons. During memory recall, the hippocampus can provide a memory index to neocortex, to selectively reinstate activity patterns across distributed neocortical networks, which manifests as a facilitatory hippocampal-neocortical interaction – here mediated polysynaptically via VIP+ and SOM+/PV+ inhibitory interneurons. Ref. 102 proposed that the diversity of inhibitory interneurons – and their selective responses to classical neuromodulators or N-methyl-D-aspartate receptor (NMDAR)-mediated stimulation–provide the necessary machinery for complementary inhibitory and facilitatory hippocampal-neocortical interactions. Computationally, the facilitatory (disinhibitory) effect of hippocampal projections would, in this scheme, encode the precision of prediction error units by modulating their postsynaptic excitability. For simplicity, we have omitted many connections and cell types in the canonical microcircuit (e.g., spiny stellate cells in layer 4) and in the hippocampus. Furthermore, we have omitted descending projections directly to PV+ interneurons. Excitatory synapses are denoted with lines ending in a circle, while inhibitory synapses are denoted by a diamond. Note that superficial pyramidal cells receive excitatory and inhibitory influences that underwrite a prediction error, while the precision of the encoded prediction error is controlled by modulatory (orange) interactions with VIP+ inhibitory interneurons. ACh refers to acetylcholine. PV, SOM and VIP refer to PV+, SOM+ and VIP+ interneurons. DG refers to the dentate gyrus, Sub refers to subiculum, which together with CA1 and CA3 constitute subfields of the hippocampus that reside along the performant pathway; "n" refers to the level in the cortical hierarchy.

Credit: Fig. 3 of ref. 102.

However vivid a dream is, we do not remember a dream unless we are woken up, and soon forget about the dream after wake-up. Cortical content that played together is likely tentatively linked (attracts hippocampal dendritic spines to grow in its direction), but weakly enough that could be easily redirected. It is also possible that information is activated before erasure; Or, information is accessed in a mode that is not possible when awake (Fig. 6.9), e.g., more defused through interneurons instead of on stable postsynaptic dendritic spines (e.g., Fig. 5.4), or a faster θ wave as one gets active. The total number of synapses decreases after REM sleep[6], freeing up surfaces for upcoming new synapses (e.g., in the next NREM), while the total number is quite constant each day (e.g., Fig. 5.7).

With local computation within dendrites (Chapter 5), results of computation would be different after such pruning of connections, which could be reestablished with more training. As with replays, things in dreams lack details such as color, while their identities are unequivocal, consistent with activity mostly at the synapses between the cortex and the hippocampus.

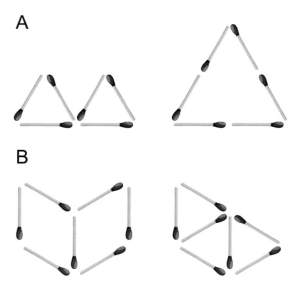

Figure 6.15 Match stick games. (A) Move two sticks to double the area. (B) Move two sticks to make three triangles.

Credit: Fei Li, Huijue Jia.

Acetylcholine potentially drives episodic activity in the neocortex (Fig. 6.10), with loosened input from the claustrum-EC-hippocampus (weaker hashing, Fig. 4.11) and from the thalamus compared to wakefulness[6] (Figs. 6.9–6.13). In contrast, acetylcholine activity in the cortex is minimum during slow-wave sleep[6], and cortical activity is likely more directed by the hippocampus (e.g., hippocampal place cells and entorhinal grid cells for space and time, Fig. 4.15 and Chapter 7). Remember the memory engrams in Chapter 4, sleeping and dreaming might be somewhat like the match sticks game we used to have in holiday homework, where limited moves are performed to reach a certain pattern (Fig. 6.15), without added burden to the brain[103].

6.6 Daydreaming and the Refreshing Effect of Switching Tasks

We cannot switch off one hemisphere of the brain like bottlenose dolphins do and continue swimming with the other hemisphere[104]. Whenever affordable, our waking brain runs some side jobs (e.g., refs. 105, 106), whether planning on the next meal, a conversation, or some other activities (replays and preplays, Chapter 5, Chapter 9). It can take too much effort to suppress such spontaneous thoughts (ripples or hotspots, Fig. 6.10) that one would rather get something done and off the mind.

With all the neural electrophysiology, the brain is traditionally believed to be dry inside during waking hours. However, slow traveling waves similar to those in sleep have been detected on a smaller scale and with lower density while awake[12, 20, 107]. I am inclined to speculate that cerebrospinal fluids also briefly go through and replenish the brain cells before returning back to blood vessels. So when we joke about having a short-circuit in the brain, there may be some truth in it. This can potentially allow some clean-up and stock-up, on a smaller scale than NREM sleep. Such a continuum of sleeping and awake will help explain the apparent lack of sleep in some animals (e.g., ref. 108). It would be too much to speculate about neurogenesis here.

When doing chores or walking, one can also become less stressed because of hormones and neurotransmitters. Locomotor speed regulates hippocampal activity (likely through parvalbumin interneurons in the entorhinal cortex[109, 110]), and hippocampal θ oscillations are stronger during movement[93, 111], likely due to modulation of θ oscillations by γ oscillations in the motor/visual cortex (Fig. 6.9) and spatial navigation (Chapter 7). Hippocampal frequency of θ rhythm is controlled by acceleration in running rats[112]. The regular breathing during walking might rhythm with the brainstem, which is known to initiate sleeping and dreaming. And the brainstem contain GABAergic interneurons that inhibit SOM interneurons in the hippocampus[113].

6.7   Summary

Sleeping is known to improve learning. The essential functions of sleep (regional instead of brain-wide in some animals) at least include a metabolic reset and a network reorganization. Things learned when awake are replayed during NREM sleep, which potentially allows new dendritic branches and new hippocampal neurons to grow with ambient supply of nutrients and signals. Stubby new dendritic spines (Chapter 5), with their more flexible orientation to catch an axon and their unrefined neck, join the competition among the more mature spines.

Sharp-wave ripples, observed when awake and asleep, look like strong candidates in milliseconds time scale that represent synaptic activity of individual dendritic spines (Chapter 5). Those that fire more get more cellular resources, including actin filaments, to sculp a larger weight (a less resistant neck[82], and perhaps also moving closer to the nucleus and decreasing resistance in the dendrite), while those that are uncertain might be unimportant enough to deserve a demolishment (forgotten, e.g., Fig. 5.7). Mechanically, the inflow of CSF and the resulting osmolarity likely allow the neurons to enlarge at hot spots, which would release their presynaptic content more easily[13], and seen as sharp-wave ripples. Hebbian learning can then get a solid physical basis.

The large-scale slower waves in NREM and REM sleep likely pumps blood and CSF flow to decrease or increase the probability of sharp-wave ripples and might normalize activity for a globally more optimal and ready-to-learn network, e.g., fewer small but strong clusters maintained by interneurons. Computers do not have to use exactly the same way, but the principles can probably help in the design of hardware and algorithms.

Questions

1. Do you try to memorize things before sleep, and does that affect your sleep? Does your brain tend to replay visual input or speeches from the day?

2. How should the variations in body temperature, acetate, etc., be incorporated into computational models of sleep-dependent memory-changes?

3. In which ways do you think the different types of interneurons work differently? How can the waves of different frequencies be modeled with a combination of excitatory and inhibitory neurons?

4. Would a neuronal path be visibly sturdier, as dendrites accumulate resources from axons and redistribute the resources to downstream or upstream regions? In that sense, even if the exact dendritic spines end up getting demolished, the network has still become more ready for something similar?

References

1. Kiviniemi, V. *et al.* Ultra-fast magnetic resonance encephalography of physiological brain activity - Glymphatic pulsation mechanisms? *J. Cereb. Blood Flow Metab.* **36**, 1033–45 (2016).

2. Rius-Pérez, S., Tormos, A. M., Pérez, S. & Taléns-Visconti, R. Vascular pathology: cause or effect in Alzheimer disease? *Neurologia* **33**, 112–120 (2018).

3. Nedergaard, M. & Goldman, S. A. Glymphatic failure as a final common pathway to dementia. *Science (80-.).* **370**, 50–56 (2020).

4. Kanekiyo, T., Liu, C.-C., Shinohara, M., Li, J. & Bu, G. LRP1 in brain vascular smooth muscle cells mediates local clearance of Alzheimer's amyloid-β. *J. Neurosci.* **32**, 16458–65 (2012).

5. Kostin, A., Alam, M. A., McGinty, D. & Alam, M. N. Adult hypothalamic neurogenesis and sleep-wake dysfunction in aging. *Sleep* **44** (2021).

6. Klinzing, J. G., Niethard, N. & Born, J. Mechanisms of systems memory consolidation during sleep. *Nat. Neurosci.* **22**, 1598–1610 (2019).

7. Xie, L. *et al.* Sleep drives metabolite clearance from the adult brain. *Science* **342**, 373–377 (2013).

8. Spano, G. M. *et al.* Sleep deprivation by exposure to novel objects increases synapse density and axon–spine interface in the hippocampal CA1 region of adolescent mice. *J. Neurosci.* **39**, 6613–6625 (2019).

9. Lewis, L. D. The interconnected causes and consequences of sleep in the brain. *Science (80-.).* **374**, 564–568 (2021).

10. Massimini, M., Huber, R., Ferrarelli, F., Hill, S. & Tononi, G. The sleep slow oscillation as a traveling wave. *J. Neurosci.* **24**, 6862–70 (2004).

11. Muller, L. *et al.* Rotating waves during human sleep spindles organize global patterns of activity that repeat precisely through the night. *Elife* **5** (2016).

12. Muller, L., Chavane, F., Reynolds, J. & Sejnowski, T. J. Cortical travelling waves: mechanisms and computational principles. *Nat. Rev. Neurosci.* **19**, 255–268 (2018).

13. Ucar, H. *et al.* Mechanical actions of dendritic-spine enlargement on presynaptic exocytosis. *Nature* **600**, 686–689 (2021).

14. Ngo, H.-V., Fell, J. & Staresina, B. Sleep spindles mediate hippocampal-neocortical coupling during long-duration ripples. *Elife* **9** (2020).

15. Girardeau, G. & Lopes-dos-Santos, V. Brain neural patterns and the memory function of sleep. *Science (80-.).* **374**, 560–564 (2021).

16. Steriade, M., Nuñez, A. & Amzica, F. A novel slow (<1 Hz) oscillation of neocortical neurons in vivo: depolarizing and hyperpolarizing components. *J. Neurosci.* **13**, 3252–65 (1993).

17. Halassa, M. M. *et al.* Selective optical drive of thalamic reticular nucleus generates thalamic bursts and cortical spindles. *Nat. Neurosci.* **14**, 1118–20 (2011).

18. Hulse, B. K., Moreaux, L. C., Lubenov, E. V & Siapas, A. G. Membrane potential dynamics of CA1 pyramidal neurons during hippocampal ripples in awake mice. *Neuron* **89**, 800–13 (2016).

19. Adamantidis, A. R., Gutierrez Herrera, C. & Gent, T. C. Oscillating circuitries in the sleeping brain. *Nat. Rev. Neurosci.* **20**, 746–762 (2019).

20. Bernardi, G. *et al.* Regional delta waves in human rapid eye movement sleep. *J. Neurosci.* **39**, 2686–2697 (2019).

21. Hasegawa, E. *et al.* Rapid eye movement sleep is initiated by basolateral amygdala dopamine signaling in mice. *Science (80-.).* **375**, 994–1000 (2022).

22. Fried, I. Neurons as will and representation. *Nat. Rev. Neurosci.* (2021) doi:10.1038/s41583-021-00543-8.

23. Kriegeskorte, N. & Wei, X.-X. Neural tuning and representational geometry. *Nat. Rev. Neurosci.* (2021) doi:10.1038/s41583-021-00502-3.

24. Uran, C. *et al.* Predictive coding of natural images by V1 firing rates and rhythmic synchronization. *Neuron* (2022) doi:10.1016/j.neuron.2022.01.002.

25. Braitenberg, V. & Schüz, A. Cortex: statistics and geometry of neuronal connectivity. *Cortex Stat. Geom. Neuronal Connect.* (1998) doi:10.1007/978-3-662-03733-1.

26. Wang, X.-J. Neurophysiological and computational principles of cortical rhythms in cognition. *Physiol. Rev.* **90**, 1195–268 (2010).

27. Kim, Y. *et al.* Brain-wide maps reveal stereotyped cell-type-based cortical architecture and subcortical sexual dimorphism. *Cell* **171**, 456-469.e22 (2017).

28. Farashahi, S. & Soltani, A. Computational mechanisms of distributed value representations and mixed learning strategies. *Nat. Commun.* **12**, 7191 (2021).

29. Campagnola, L. *et al.* Local connectivity and synaptic dynamics in mouse and human neocortex. *Science (80-.).* **375**, (2022).

30. Buzsáki, G. Hippocampal sharp wave-ripple: a cognitive biomarker for episodic memory and planning. *Hippocampus* **25**, 1073–188 (2015).

31. Joffe, M. E. *et al.* Acute restraint stress redirects prefrontal cortex circuit function through mGlu5 receptor plasticity on somatostatin-expressing interneurons. *Neuron* (2022) doi:10.1016/j.neuron.2021.12.027.

32. Huang, Z. J. & Paul, A. The diversity of GABAergic neurons and neural communication elements. *Nat. Rev. Neurosci.* **20**, 563–572 (2019).

33. Bugeon, S. *et al.* A transcriptomic axis predicts state modulation of cortical interneurons. *Nature* (2022) doi:10.1038/s41586-022-04915-7.

34. Markram, H. *et al.* Interneurons of the neocortical inhibitory system. *Nat. Rev. Neurosci.* **5**, 793–807 (2004).

35. Kubota, Y. *et al.* Functional effects of distinct innervation styles of pyramidal cells by fast spiking cortical interneurons. *Elife* **4** (2015).

36. Hu, H., Martina, M. & Jonas, P. Dendritic mechanisms underlying rapid synaptic activation of fast-spiking hippocampal interneurons. *Science (80-.).* **327**, 52–58 (2010).

37. Topolnik, L. & Tamboli, S. The role of inhibitory circuits in hippocampal memory processing. *Nat. Rev. Neurosci.* (2022) doi:10.1038/s41583-022-00599-0.

38. Mukherjee, A., Lam, N. H., Wimmer, R. D. & Halassa, M. M. Thalamic circuits for independent control of prefrontal signal and noise. *Nature* **600**, 100–104 (2021).

39. Roy, D. S., Zhang, Y., Halassa, M. M. & Feng, G. Thalamic subnetworks as units of function. *Nat. Neurosci.* **25**, 140–153 (2022).

40. Kim, S., Wallace, M. L., El-Rifai, M., Knudsen, A. R. & Sabatini, B. L. Co-packaging of opposing neurotransmitters in individual synaptic vesicles in the central nervous system. *Neuron* (2022) doi:10.1016/j.neuron.2022.01.007.

41. Cossart, R. & Garel, S. Step by step: cells with multiple functions in cortical circuit assembly. *Nat. Rev. Neurosci.* (2022) doi:10.1038/s41583-022-00585-6.

42. Blackman, A. V., Abrahamsson, T., Costa, R. P., Lalanne, T. & Sjöström, P. J. Target-cell-specific short-term plasticity in local circuits. *Front. Synaptic Neurosci.* **5** (2013).

43. Perrault, A. A. *et al.* Whole-night continuous rocking entrains spontaneous neural oscillations with benefits for sleep and memory. *Curr. Biol.* **29**, 402-411.e3 (2019).

44. Tripathy, S. J., Burton, S. D., Geramita, M., Gerkin, R. C. & Urban, N. N. Brain-wide analysis of electrophysiological diversity yields novel categorization of mammalian neuron types. *J. Neurophysiol.* **113**, 3474–3489 (2015).

45. K, T. *et al.* Sleep and second-language acquisition revisited: the role of sleep spindles and rapid eye movements. *Nat. Sci. Sleep* **13**, 1887–1902 (2021).

46. E, C. *et al.* Sleep spindles promote the restructuring of memory representations in ventromedial prefrontal cortex through enhanced hippocampal-cortical functional connectivity. *J. Neurosci.* **40**, 1909–1919 (2020).

47. Hablitz, L. M. *et al.* Increased glymphatic influx is correlated with high EEG delta power and low heart rate in mice under anesthesia. *Sci. Adv.* **5**, eaav5447 (2019).

48. Zelano, C. *et al.* Nasal respiration entrains human limbic oscillations and modulates cognitive function. *J. Neurosci.* **36**, 12448–12467 (2016).

49. Tseng, Y.-T. *et al.* The subthalamic corticotropin-releasing hormone neurons mediate adaptive REM-sleep responses to threat. *Neuron* (2022) doi:10.1016/j.neuron.2021.12.033.

50. Baehr, E. K., Revelle, W. & Eastman, C. I. Individual differences in the phase and amplitude of the human circadian temperature rhythm: with an emphasis on morningness-eveningness. *J. Sleep Res.* **9**, 117–127 (2000).

51. Buhr, E. D., Yoo, S.-H. & Takahashi, J. S. Temperature as a universal resetting cue for mammalian circadian oscillators. *Science* **330**, 379–385 (2010).

52. Jie, Z. *et al.* A transomic cohort as a reference point for promoting a healthy human gut microbiome. *Med. Microecol.* **8**, 100039 (2021).

53. Leone, V. *et al.* Effects of diurnal variation of gut microbes and high-fat feeding on host circadian clock function and metabolism. *Cell Host Microbe* **17**, 681–689 (2015).

54. Jia, H. *Investigating Human Diseases with the Microbiome: Metagenomics Bench to Bedside.* (Elsevier, 2022).

55. Szentirmai, É., Millican, N. S., Massie, A. R. & Kapás, L. Butyrate, a metabolite of intestinal bacteria, enhances sleep. *Sci. Rep.* **9**, 7035 (2019).

56. Harada, N. *et al.* Hypogonadism alters cecal and fecal microbiota in male mice. *Gut Microbes* **7**, 533–539 (2016).

57. Xiong, J. *et al.* FSH blockade improves cognition in mice with Alzheimer's disease. *Nature* 1–7 (2022) doi:10.1038/s41586-022-04463-0.

58. Kornman, K. S. & Loesche, W. J. Effects of estradiol and progesterone on Bacteroides melaninogenicus and Bacteroides gingivalis. *Infect. Immun.* **35**, 256–263 (1982).

59. Ahmed, S. M. H. *et al.* Fitness trade-offs incurred by ovary-to-gut steroid signalling in Drosophila. *Nature* (2020) doi:10.1038/s41586-020-2462-y.

60. Chen, C. *et al.* Cervicovaginal microbiome dynamics after taking oral probiotics. *J. Genet. Genomics* (2021) doi:10.1016/j.jgg.2021.03.019.

61. Jie, Z. *et al.* Life history recorded in the vagino-cervical microbiome along with multi-omics. *Genomics. Proteomics Bioinformatics* (2021) doi:10.1016/j.gpb.2021.01.005.

62. Jie, Z. *et al.* Dairy consumption and physical fitness tests associated with fecal microbiome in a Chinese cohort. *Med. Microecol.* 100038 (2021) doi:10.1016/j.medmic.2021.100038.

63. Shein-Idelson, M., Ondracek, J. M., Liaw, H. P., Reiter, S. & Laurent, G. Slow waves, sharp waves, ripples, and REM in sleeping dragons. *Science (80-.).* **352**, 590–595 (2016).

64. Baden, T., Euler, T. & Berens, P. Understanding the retinal basis of vision across species. *Nat. Rev. Neurosci.* **21**, 5–20 (2020).

65. Zhu, F. *et al.* Transplantation of microbiota from drug-free patients with schizophrenia causes schizophrenia-like abnormal behaviors and dysregulated kynurenine metabolism in mice. *Mol. Psychiatry* (2019) doi:10.1038/s41380-019-0475-4.

66. Zhu, F. *et al.* Metagenome-wide association of gut microbiome features for schizophrenia. *Nat. Commun.* **11**, 1612 (2020).

67. Zheng, P. *et al.* The gut microbiome from patients with schizophrenia modulates the glutamate-glutamine-GABA cycle and schizophrenia-relevant behaviors in mice. *Sci. Adv.* **5**, eaau8317 (2019).

68. Medeiros, S. L. de S. *et al.* Cyclic alternation of quiet and active sleep states in the octopus. *iScience* **24**, 102223 (2021).

69. Frank, M. G., Waldrop, R. H., Dumoulin, M., Aton, S. & Boal, J. G. A preliminary analysis of sleep-like states in the cuttlefish sepia officinalis. *PLoS One* **7**, e38125 (2012).

70. Nir, Y. & Tononi, G. Dreaming and the brain: from phenomenology to neurophysiology. *Trends Cogn. Sci.* **14**, 88–100 (2010).

71. Siclari, F. *et al.* The neural correlates of dreaming. *Nat. Neurosci.* **20**, 872–878 (2017).

72. Yang, Y. *et al.* Theta-gamma coupling emerges from spatially heterogeneous cholinergic neuromodulation. *PLoS Comput. Biol.* **17** (2021).

73. Diekelmann, S. & Born, J. The memory function of sleep. *Nat. Rev. Neurosci.* **11**, 114–126 (2010).

74. W, L., L, M., G, Y. & WB, G. REM sleep selectively prunes and maintains new synapses in development and learning. *Nat. Neurosci.* **20**, 427–437 (2017).

75. T, G., L, G., DS, R., A, B. & K, G. Neural reactivations during sleep determine network credit assignment. *Nat. Neurosci.* **20**, 1277–1284 (2017).

76. AJ, H., JA, B., E, Z., D, B. & N, B. Grid-like processing of imagined navigation. *Curr. Biol.* **26**, 842–847 (2016).

77. Yao, Z. *et al.* Abnormal cortical networks in mild cognitive impairment and Alzheimer's disease. *PLoS Comput. Biol.* **6**, e1001006 (2010).

78. JB, A. & MC, C. Role of emergent neural activity in visual map development. *Curr. Opin. Neurobiol.* **24**, 166–175 (2014).

79. Roscow, E. L., Chua, R., Costa, R. P., Jones, M. W. & Lepora, N. Learning offline: memory replay in biological and artificial reinforcement learning. *Trends Neurosci.* **44**, 808–821 (2021).

80. de Vivo, L. *et al.* Ultrastructural evidence for synaptic scaling across the wake/sleep cycle. *Science* **355**, 507–510 (2017).

81. D, K. *et al.* Sparse activity of hippocampal adult-born neurons during REM sleep is necessary for memory consolidation. *Neuron* **107**, 552-565.e10 (2020).

82. Crick, F. Do dendritic spines twitch? *Trends Neurosci.* **5**, 44–46 (1982).

83. Berry, K. P. & Nedivi, E. Spine dynamics: are they all the same? *Neuron* **96**, 43–55 (2017).

84. Frank, A. C. *et al.* Hotspots of dendritic spine turnover facilitate clustered spine addition and learning and memory. *Nat. Commun.* **9** (2018).

85. Nägerl, U. V., Köstinger, G., Anderson, J. C., Martin, K. A. C. & Bonhoeffer, T. Protracted synaptogenesis after activity-dependent spinogenesis in hippocampal neurons. *J. Neurosci.* **27**, 8149–56 (2007).

86. Cobb, S. R., Buhl, E. H., Halasy, K., Paulsen, O. & Somogyi, P. Synchronization of neuronal activity in hippocampus by individual GABAergic interneurons. *Nature* **378**, 75–78 (1995).

87. Staudigl, T. *et al.* Hexadirectional modulation of high-frequency electrophysiological activity in the human anterior medial temporal lobe maps visual space. *Curr. Biol.* **28**, 3325-3329.e4 (2018).

88. Vaz, A. P., Inati, S. K., Brunel, N. & Zaghloul, K. A. Coupled ripple oscillations between the medial temporal lobe and neocortex retrieve human memory. *Science* **363**, 975–978 (2019).

89. Norman, Y. *et al.* Hippocampal sharp-wave ripples linked to visual episodic recollection in humans. *Science* **365** (2019).

90. Abbaspoor, S., Hussin, A. T. & Hoffman, K. L. Theta- and gamma-band oscillatory uncoupling in the macaque hippocampus. *bioRxiv* 2021.12.30.474585 (2022) doi:10.1101/2021.12.30.474585.

91. Sarter, M. & Lustig, C. Forebrain cholinergic signaling: wired and Phasic, Not tonic, and causing behavior. *J. Neurosci.* **40**, 712–719 (2020).

92. Sirota, A. *et al.* Entrainment of neocortical neurons and gamma oscillations by the hippocampal theta rhythm. *Neuron* **60**, 683–697 (2008).

93. Bush, D. *et al.* Human hippocampal theta power indicates movement onset and distance travelled. *Proc. Natl. Acad. Sci.* **114**, 12297–12302 (2017).

94. Oberto, V. J. *et al.* Distributed cell assemblies spanning prefrontal cortex and striatum. *Curr. Biol.* (2021) doi:10.1016/j.cub.2021.10.007.

95. Lubenov, E. V & Siapas, A. G. Hippocampal theta oscillations are travelling waves. *Nature* **459**, 534–539 (2009).

96. Patel, J., Fujisawa, S., Berényi, A., Royer, S. & Buzsáki, G. Traveling theta waves along the entire septotemporal axis of the hippocampus. *Neuron* **75**, 410–417 (2012).

97. Jung, M. W., Wiener, S. I. & McNaughton, B. L. Comparison of spatial firing characteristics of units in dorsal and ventral hippocampus of the rat. *J. Neurosci.* **14**, 7347–7356 (1994).

98. Kjelstrup, K. B. *et al.* Finite scale of spatial representation in the hippocampus. *Science* **321**, 140–143 (2008).

99. Tort, A. B. L., Scheffer-Teixeira, R., Souza, B. C., Draguhn, A. & Brankačk, J. Theta-associated high-frequency oscillations (110–160 Hz) in the hippocampus and neocortex. *Prog. Neurobiol.* **100**, 1–14 (2013).

100. Echeveste, R., Aitchison, L., Hennequin, G. & Lengyel, M. Cortical-like dynamics in recurrent circuits optimized for sampling-based probabilistic inference. *Nat. Neurosci.* **23**, 1138–1149 (2020).

101. PA, L. & D, B. How Targeted memory reactivation promotes the selective strengthening of memories in sleep. *Curr. Biol.* **29**, R906–R912 (2019).

102. Barron, H. C., Auksztulewicz, R. & Friston, K. Prediction and memory: a predictive coding account. *Prog. Neurobiol.* **192**, 101821 (2020).

103. Bullmore, E. & Sporns, O. The economy of brain network organization. *Nat. Rev. Neurosci.* **13**, 336–349 (2012).

104. Siclari, F. & Tononi, G. Local aspects of sleep and wakefulness. *Curr. Opin. Neurobiol.* **44**, 222–227 (2017).

105. Musall, S., Kaufman, M. T., Juavinett, A. L., Gluf, S. & Churchland, A. K. Single-trial neural dynamics are dominated by richly varied movements. *Nat. Neurosci.* **22**, 1677–1686 (2019).

106. Ólafsdóttir, H. F., Carpenter, F. & Barry, C. Task demands predict a dynamic switch in the content of awake hippocampal replay. *Neuron* **96**, 925-935.e6 (2017).

107. VV, V. *et al.* Local sleep in awake rats. *Nature* **472**, 443–447 (2011).

108. Cirelli, C. & Tononi, G. Is sleep essential? *PLoS Biol.* **6**, e216 (2008).

109. Miao, C., Cao, Q., Moser, M.-B. & Moser, E. I. Parvalbumin and somatostatin interneurons control different space-coding networks in the medial entorhinal cortex. *Cell* **171**, 507-521.e17 (2017).

110. Dannenberg, H., Lazaro, H., Nambiar, P., Hoyland, A. & Hasselmo, M. E. Effects of visual inputs on neural dynamics for coding of location and running speed in medial entorhinal cortex. *Elife* **9**, 1–34 (2020).

111. Farrell, J. S. *et al.* Supramammillary regulation of locomotion and hippocampal activity. *Science (80-.).* **374**, 1492–1496 (2021).

112. Kropff, E., Carmichael, J. E., Moser, E. I. & Moser, M.-B. Frequency of theta rhythm is controlled by acceleration, but not speed, in running rats. *Neuron* **109**, 1029-1039.e8 (2021).

113. Szőnyi, A. *et al.* Brainstem nucleus incertus controls contextual memory formation. *Science (80-.).* **364** (2019).

Chapter 7

Mastering Space and Time

Abstract

In our brain, grid cells, object vector cells, head direction cells, speed cells and other cells work together with hippocampal place cells for an accurate representation of space, and possibly time. Inaccuracies in direction, distance and place can result in loss of crucial opportunities. Memory and learning (Chapters 4–6) also work in this framework of space and time. The brain probably has a better Generative Adversarial Network (GAN) than state-of-the-art algorithms. In addition, social navigation takes place in two-dimensional grids and vectors as well. We'll see more abstract and generative use of this navigation system in the subsequent Chapters (Chapters 8, 9) for higher level cognition.

Keywords

Hippocampus, Medial entorhinal cortex (MEC), Stellate cells, Grid cells, Spatial navigation, Goal-directed navigation, Generative Adversarial Networks (GAN), Claustrum, Attention, Social navigation

7.1 Place Cells and Grid Cells

The spatial navigation systems in the brain, e.g., head direction cells, start developing long before an animal opens its eyes for the

Neuroscience for Artificial Intelligence
Huijue Jia
Copyright © 2023 Jenny Stanford Publishing Pte. Ltd.
ISBN 978-981-4968-78-2 (Hardcover), 978-1-003-41098-0 (eBook)
www.jennystanford.com

first time. We already heard about hippocampal place cells and their replays together with cortical neurons in Chapters 4 and 5. In addition to a gradient around familiar places, places or other items and people remembered need to be organized, especially when trying to move far from a familiar area and walk around in a new place. More than 10% of cells in the medial entorhinal cortex (MEC, Figs. 1.6 and 4.9) clearly show a hexagonal grid-like firing pattern, and have thus been called grid cells[1, 2] (Figs. 7.1 and 7.2).

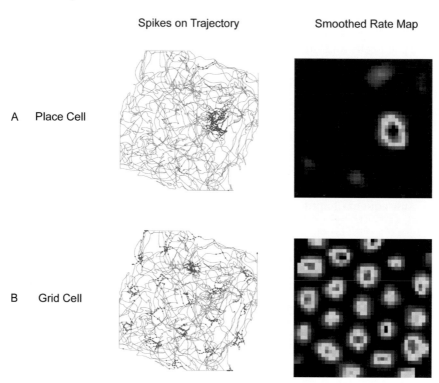

Figure 7.1 Firing fields of place cells in the hippocampus (A) and grid cells in the medial entorhinal cortex (MEC) (B). The first column shows the spike locations (red dots) of the cells superimposed on the trajectory of the animal during a trial. The second and third columns show the smoothed spatial rate maps of the cells. Color coding from blue (min.) to red (max.) is used for each spiking rate map. Whereas most place cells have a single firing location, the firing fields of a grid cell form a periodic triangular matrix tiling the entire environment available to the animal.

Credit: Left and right rows of Fig. 2B,C of ref. 3.

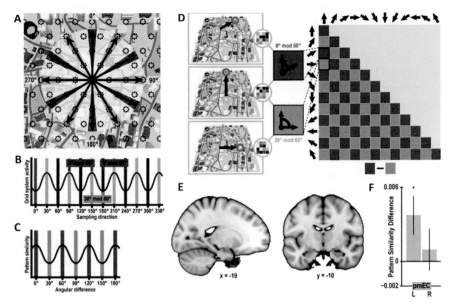

Figure 7.2 Grid-like representations during imagined navigation. (A) Six-fold symmetric firing fields of a hypothetical grid cell (dark blue dotted circles) superimposed on an aerial view of the virtual city Donderstown, which the volunteers have been trained to navigate in. Black arrows indicate the twelve sampled directions; light and dark shading highlights directions (multiples of) 60° apart. For illustration purposes, the grid orientation is aligned to the sampled directions; other example in ref. 9. (B) The firing rate of the hypothetical response of the grid cell system as a function of direction, showing a 60° modulation. Shading displays sampling of directions and red and blue markers indicate the two conditions. Note that the oscillatory firing pattern is sampled at the same phase in the 0° modulo 60° condition, but at different phases in the 30° modulo 60° condition. (C) Based on this, the authors of ref. 9 expected a 60° modulation of fMRI pattern similarity values when comparing trial pairs based on the angular difference of their sampled directions. Red and blue shading illustrates the two conditions. (D) Specifically, the authors predicted higher pattern similarity for trial pairs with a remainder of 0° (0° modulo 60° condition, red) compared to trial pairs with a remainder of 30° (30° modulo 60° condition, blue), when dividing the angular difference of the pair's sampling directions by 60°. Note that for illustration purposes the predicted similarity matrix is shown for comparisons across conditions, not single trials. (E) Region of interest (ROI) mask for posterior medial entorhinal cortex (pmEC) from previous report[10]. (F) Pattern similarity difference (mean and SEM) between the two conditions. The left pmEC exhibited a significant 60° modulation of pattern similarity. No significant differences in pattern similarity were observed in the right pmEC (T_{23} = 0.57, p = 0.58).

Credit: Fig. 3 of ref. 9.

In Alzheimer's patients, accumulation of tau (a protein that forms filaments and can be toxic to neurons) is first seen in the entorhinal cortex, and over time, the hippocampus and then the cortex (Layer III first) are also affected[4]. Young adults carrying genetic risk for Alzheimer's disease (the *APOE*-ε4 allele) already showed reduced grid-like representation detectable by fMRI (functional Magnetic Resonance Imaging) and impaired spatial memory performance[5]. On the other hand, their hippocampus was more active, possibly as a compensate[5], and consistent with better visual working memory of the *APOE*-ε4 carriers[6]. Grid cell spatial representation, synchrony with interneurons and head direction cells were impaired in a transgenic mice model that overexpress amyloid β protein (Aβ, mentioned for sleep in Section 6.1) in the hippocampus and entorhinal cortex[7]. In patients with major depressive disorder, history of childhood physical or emotional trauma negatively correlated with entorhinal thickness, while positive personality traits correlated with larger volumes of hippocampus and amygdala and thicker cortex in the precuneus and cingulate cortex[8].

Grid cells in the MEC have hexagonal grid-like firing fields that work together with place fields of the hippocampal pyramidal neurons (place cells) (Figs. 7.1, 7.2, 7.3 and 7.4)[11-15]. Even in an open field, the grids are tilted relative to the perceived borders, so that each place is more unique (more of a problem for a round field).

Like the increasing size of place fields from the dorsal to the ventral hippocampus (posterior-anterior in humans, which is also the direction of θ wave phase propagation, Fig. 6.12), the grids were reported to increase in size along the dorsal-ventral axis, and differ in orientation[12, 15, 16]. But there are more cells with smaller grid scales; and on the ventral end, cells with the largest grid scales appeared to coexist with cells of smaller grid scales[17] (Fig. 7.5; Section 7.2 for stellate cells). During development, the MEC layers mature from the dorsal to the ventral end in mice[18], which would be posterior-anterior in humans (consistent with Section 3.2).

Grid cells in rats can be clustered into about five modules, and the few individual animals studied already showed different grid sizes[17, 19] (Fig. 7.5). Consistent with theoretical calculation, the grid scales of successive modules showed a ratio of no more than 1.5[17, 19, 20]. Skipping one or two modules resulted in larger ratios (Fig. 7.5C), which could lead to catastrophic error in defining a position (Fig. 7.3), or generate stochasity[21].

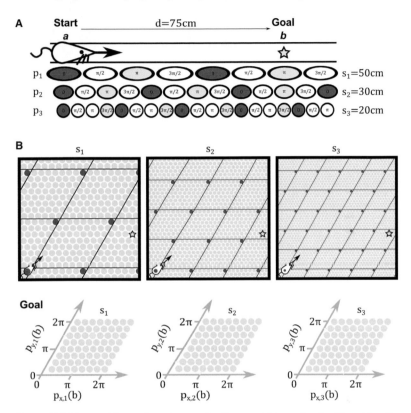

Figure 7.3 The problem of vector navigation could be solved with grid cells. (A) In 1D space, we can visualize each module of grid cells as a ring that supports a population activity bump centered at phase p_i where $0 \leq p_i < 2\pi$. The displacement d between the starting location a (red) and goal location b (yellow) can be calculated from the grid cell representations of those locations (i.e., sets of spatial phases across grid modules $\{p_i(a)\}$ and $\{p_i(b)\}$). (B) In 2D space, we can visualize each module of grid cells as a twisted torus supporting a single activity bump centered at phases $p_i \rightarrow = (p_{x,i}, p_{y,i})$ along the principal axes of a unitary "tile" of the grid pattern (i.e., unit vectors $x\rightarrow$ and $y\rightarrow$). The distance and direction between start and goal locations can be defined by the grid cell representations of those locations (i.e., sets of spatial phases across grid modules and principal axes: $\{p_x(a\rightarrow), p_y(a\rightarrow)\}, \{p_x(b\rightarrow), p_y(b\rightarrow)\}\{p_x(b\rightarrow), p_y(b\rightarrow)\}$).

Credit: Fig. 3 of ref. 24.

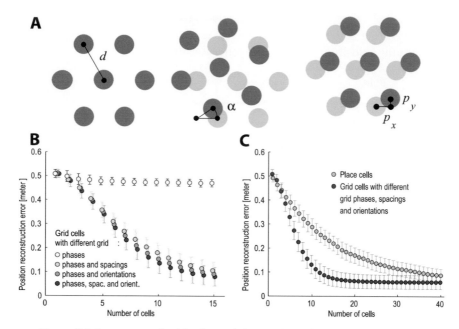

Figure 7.4 Parameters of grid cells, and the number of cells needed to define a position. (A) Characterization of the grids with three parameters: grid spacing d, grid orientation α and grid phase $p = (p_x, p_y)$ (along the x- and y-axes). Circles represent grid subfields. (B) Comparison of the position reconstruction error of four different subsets of simulated grid cells: grid cells with different grid phases, with different grid phases and spacings, with different grid phases and orientations and with different grid phases, spacings and orientations. Error bars indicate ± s.d. (C) Comparison of the position reconstruction error of a set of simulated grid cells with different grid phases, spacings and orientations with a set of simulated place cells, as a function of the number of cells.

Credit: Part A, Fig. 1a,b,c of ref. 16; Part B and C, Fig. 4c,d of ref. 16.

Activity in a single module is consistent with a torus[19] (donut shape, Fig. 8.5), which would possibly allow more tweaking than a flat grid. Rat grid cells in the same modules showed similar orientation and similar intrinsic θ frequency[17] (interneurons in Chapter 6). θ waves and phases in grid cells, in relation to those in place cells, have been proposed to signal the direction of movement[22, 23].

Combination of only a few grid cells of different grid sizes would be sufficient to uniquely define a target place (Fig. 7.3)[21, 24]. The phases of grid cell firing have to combine with grid spacing and grid orientation to accurately reconstruct a place[16] (Fig. 7.4, more advanced use in Chapter 9). The size of the firing field should not be too small relative to the grid cell spacing. Interestingly, the same number of grid cells are more efficient than the same number of place cells in defining a position[16] (Fig. 7.4)—which likely holds true for higher level cognition (Chapters 8, 9). At the spot, individual cells of place cells have a smaller position reconstruction error than grid cells[16].

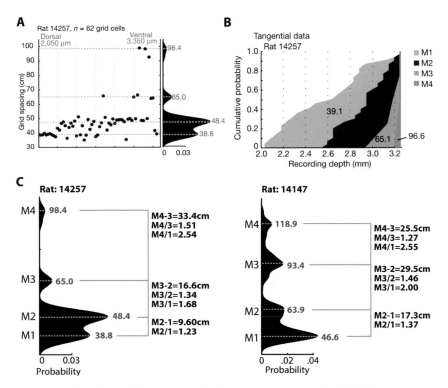

Figure 7.5 Grid modules are organized as overlapping bands. (A) Grid spacing at successive dorsoventral levels in a single rat (TT, tetrode). Dots correspond to individual cells. Cells are plotted sequentially to avoid overlap between cells at similar depths. Right, kernel smoothed density (KSD) estimate of the distribution. Red text, spacing in cm for the estimated peaks. (B) Cumulative distribution of grid modules as a function of dorsoventral position in rat 14257 (62 grid cells, k-means-determined modules are color coded). Mean grid spacings in cm are indicated. (C) Kernel smoothed density estimates of grid spacing from two representative animals with four modules. All measures (scale increments, ratios and normalized scale) reveal highly irregular relationships between successive module peaks for grid scale (green dashed lines, peak values [cm] shown in red).

Credit: Part A, Fig. 1d; Part B, Fig. 4a; Part C, Sup. Fig. 18a of ref. 17.

Between ~6–15 cells, the triangular grids are more accurate than square or hexagonal grids, with maximum number of cells in a given area[16]. The triangular arrangement, which can be generated with a twisted torus (donut shape, twisted so that the grids are not rectangular, Fig. 8.5), provides the densest possible grid for robust calibration with overlapping place cells[25].

MEC cells with band-like firing patterns are not fundamentally different from grid cells[26, 27] (Fig. 7.6). Bats were reported to have some three-dimensional (3D) grid cells in addition to two-dimensional (2D) grid cells[28], whose place fields likely depend on perception of the 2D plane, and shearing that distort the grids[15, 29]. Bat pups first learned to navigate when being carried upside-down by their mothers[30].

Although with a very different brain structure (Chapter 1, Figs. 1.1 and 1.2), fish also has an internal map of space that can be either egocentric (body-centered) or allocentric (world-centered, e.g., landmarks); and effectively make shortcuts or detours to goal locations at the first trial[31]—unlike existing AI algorithms that require extensive training data and still lack flexibility.

7.2 Stellate Neurons and Pyramidal Neurons for Objects and Grids?

Pyramidal neurons in MEC layer II (L2) form synchronously firing clusters arranged in hexagonal grids reminiscent of grid cells[13] and torus. Only the pyramidal cells at the edge of each cluster showed a high grid score[32], and each cluster can contain 800 pyramidal cells in humans[20, 33]. The pyramidal neurons are phase-locked with hippocampal θ waves[13].

Spiny stellate cells* (excitatory interneurons[34], Fig. 1.9) in MEC L2, on the other hand, form a uniform lattice (Figs. 7.6 and 7.14 in Section 7.8), like pixels of an image. The stellate cells mature earlier than pyramidal cells in MEC L2 and hippocampal CA3 during development[18]. Stellate cells can represent borders[35]. Grids

---

*The term "stellate cell" is inconsistently used in the literature and more consensus is needed. Dr. Almut Schüz only uses the term 'stellate cells' for non-spiny, inhibitory neurons, while for many authors the term includes both excitatory and inhibitory neurons.

appear distorted at the borders and by objects[14, 17, 36, 37]. My current guess is that some stellate neurons, as allocentric reference points (like pixels), represent object vector cells[1, 13, 20, 32, 38]; Stellate cells that overlap with the cluster of pyramidal neurons show up as grid cells (Figs. 7.6 and 7.5), and might incorporate other information such as head direction[19, 39, 40]. Parvalbumin-expressing interneurons (Figs. 6.6 and 6.7) help maintain the grid pattern[41], and are more modulated by speed than the other neurons[41, 42].

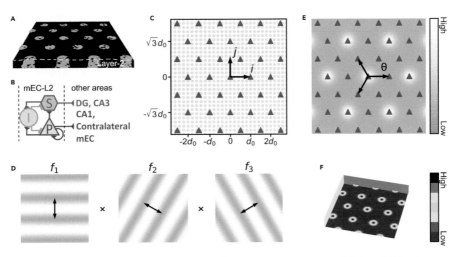

Figure 7.6 Schematic of the model[20], activation pattern and firing fields of grid cells. (A) Schematic illustration of principal neurons in mEC-L2. Stellate cells are uniformly distributed, while pyramidal cells gather in patches distributed hexagonally (green). (B) Schematically shown are connections between different neural types. S: stellate cell; P: pyramidal cell; I: interneuron. (C) Each module includes 39 hexagonally distributed pyramidal patches (green triangles) and uniformly distributed stellate cells (gray dots) besides interneurons. Each neuron is labeled by its location pp in the i, j-coordinate, where the i-axis parallels the dorsal-ventral axis of mEC-L2. Note that the stellate cells are likely also arranged in triangular grids, as would be theoretically expected for the maximal number of overlaps[16, 25]. (D) Activation state of a cell is described by the product of three von Mises functions with distinct orientations. The arrow in each panel denotes the orientation of each function. (E) Pattern of activity bumps (white) and pyramidal patches (green). θ labels the pattern orientation. (F) Shown are the firing fields of one grid cell, distributed hexagonally, when an animal ran in a square box. The firing rate is color coded.

Credit: Fig. 1 of ref. 20.

Visual contrast detected by the primary visual cortex (V1, Chapter 2) allow object vector cells to track objects (projected on 2D) regardless of size or shape, and estimate the animal's direction and distance from the objects[43–45]. Edges and patterns detected in vision might also guide and prioritize the placing of grids (including during reading, Section 8.7.2).

Stellate cells in MEC L2 have an apparent firing field smaller than that of pyramidal cells[46], while having a more extended dendritic tree[13] and a larger cell body[32]. The object vector cells are expected to complement the more egocentric grid cells (likely coupled with cortical γ oscillations and hippocampal θ waves), which remap if not well-aligned[37, 45]. Stellate cells (excitatory interneurons[34]) and inhibitory interneurons might also tweak the torus of grid firing (which might be represented by the edge of a cluster of pyramidal cells) with information on the direction and speed of movement.

7.3 Time or Rhythm?

"Im Sturz durch Raum und Zeit, Richtung Unendlichkeit..." (During the fall through space and time, towards infinity...), goes the German band Nena's song *Irgendwie, irgendwo, irgendwann*. In our brain, time does probably go together with space. Hippocampal neurons that fire at successive moments (while also encoding location and behavior) have been referred to as "time cells"[47]. Whether according to time or distance[48] (Fig. 7.7), it looks like the rats expect to finish the treadmill run anyway. Again, the moments would need to be encoded and played in sequence, and we saw plenty of waves in Chapter 6 on sleep.

The MEC[1] and the hippocampus (e.g., ref. 49) contribute to the (time frame and) temporal sequence of memory (e.g., Fig. 4.15). Moreover, the anterior-lateral entorhinal cortex (alEC) in humans (lateral entorhinal cortex (LEC) in rodents, a major source of cortical input to the hippocampus) plays a role in recalling the temporal order of events, i.e., episodic time[50, 51]. LEC dominates the EC input on new neurons in the hippocampal dentate gyrus (Fig. 5.14).

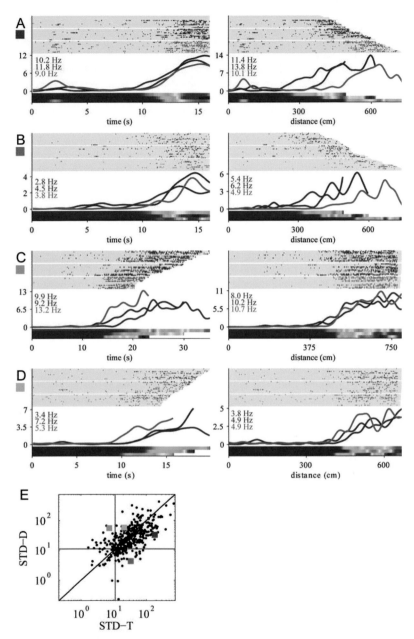

Figure 7.7 Hippocampal coding for time and distance. Examples of two cells that were more strongly influenced by time (A and B) and two cells that were more strongly influenced by distance (C and D). For each neuron, the
(*Continued*)

Figure 7.7 *(Continued)*

same firing activity is plotted both as a function of time since the treadmill started (left panels) and distance traveled on the treadmill (right panels). Blue, brown, and green ticks (and tuning curves) represent the slowest 1/3 of runs, middle 1/3 of runs, and fastest 1/3 of runs, respectively. On each trial, the rats entered the central stem of the maze from one of two directions (left or right), and then walked onto the treadmill where they received a small water reward. After a short delay, the treadmill accelerated to a speed randomly chosen from within a predetermined range, and the rats ran in place until the treadmill stopped automatically and another small water reward was delivered. Subsequently the animals finished the trial by turning in the direction opposite from their entry to the stem (spatial alternation) to arrive at a water port at the end of a goal arm. Numbers in blue, brown, and green indicate the peak firing rate in spikes per second (Hz) of the corresponding group of runs. The rows in the raster plots represent treadmill runs sorted in order of slowest speed (on top) to fastest speed (on bottom). Colored squares to the left edge of each neuron (panels A–D) Correspond to colored squares in panel E. A Generalized Linear Model (GLM) was used to quantify the effects of space (position), time and distance, and found 11% of the neurons to show only distance as informative and 8% of the neurons to show only time as informative (E). x values are the deviances of the space and distance ("S+D") model from the full ("S+T+D") model (the result of removing time from the full model, hence the label "STD-T"). A larger x value indicates a more significant contribution from time. y values are the deviances of the space and time ("S+T") model from the full ("S+T+D") model (the result of removing distance from the full model, hence the label "STD-D"). A larger y value indicates a more significant contribution from distance. Each point represents a single neuron. The red lines indicate the minimum thresholds for significance. Points in the upper-right quadrant had a significant influence of both distance and time. Points in the upper-left quadrant had a significant influence of just distance. Points in the lower-right quadrant had a significant influence of just time. Points in the lower-left quadrant were not significantly influenced by either time or distance. Points along the diagonal have an equal contribution from distance and time.

Credit: Parts A-D, Fig. 7 of ref. 48; Part E, Fig. 8A of ref. 48.

At much longer time scales, a radical idea with adult neurogenesis and sleep (Chapters 4 and 6) would be that, with sparse threads kept with the EC and with the cortex, it is possible that some of the nodes in the hippocampus (Figs. 5.14, 4.9 and 1.6) gradually move, which contribute to the decrease in temporal resolution[52]. That would also allow one to search for and repair information within a time window. There is some evidence that the anterior hippocampus responds to novelty and posterior hippocampus responds to previously encountered stimuli[52]. Cued by photos from a custom lifelogging device, the left anterior hippocampus was found to represent space and time for a month of remembered events occurring over distances of up to 30 km[53], while the posterior hippocampus is known for finer separation[54].

In rat experiments, drugs against anxiety lowered the intercept (resting level) of the linear relationship between running speed and the frequency of hippocampal θ oscillations; Environmental novelty decreased the slope of the line, likely due to the larger grids in use for novel environments which also expands place cell firing fields[55]. We enjoy focused activities that make us forget about time, perhaps indeed because the hippocampal-cortical oscillations are stable and we don't have to switch cells too often.

7.4   Sensing Speed and Acceleration

Speed may be a universal excitement. From the pedunculopontine tegmental nucleus of the brainstem, the horizontal limb of the diagonal band of Broca, to the MEC, there are neurons that increase firing with higher speed and there are neurons that decrease firing with higher speed[42, 56], which looks like a robust system as a life-or-death function should be. Inhibition of parvalbumin-expressing interneurons (corresponding to basket cells or chandelier cells[34, 57], Figs. 6.6 and 6.7) impairs speed tuning[41, 42].

With speed cells in the MEC, the hippocampus appears to complement the system with information on acceleration. Hippocampal frequency of θ rhythm (which scales with the excitatory current input[58]) is controlled by acceleration in running rats, further adding to the accuracy of navigation[59] (more on θ waves in Section 7.8). Interestingly, high-intensity interval training (HIIT) instead of jogging is recommended for people with preclinical type 2 diabetes or Alzheimer's disease[60, 61].

7.5   The Vestibular System for Sensing Self-Motion

The three semicircular canals in the inner ear make it possible for us to sense self-motion in three-dimensional space (Figs. 7.8

and 2.5). The fluids disturb bundles of "hair" on the hair cells, and the signals are integrated for the perception of self-motion (Fig. 7.8). With some training, the vestibular cerebellar neurons in a monkey sitting on a rotating platform learned not to fire and would fire again if the rotation stops (Fig. 7.9), perhaps dizzy like us when our spinning chair has been stopped.

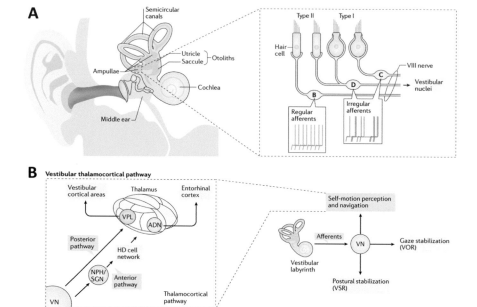

Figure 7.8 Overview of the vestibular labyrinth and central pathways. (A) The vestibular labyrinth comprises five end organs: the three semicircular canals and the two otoliths (utricle and the saccule). In mammals, there are two types of hair cell within each of the vestibular sensory organs: cylindrical type II hair cells and the phylogenetically older flask-shaped type I hair cells. Peripheral afferents in cranial nerve VIII innervate hair cells and carry head movement signals to the vestibular nuclei and to some regions of the vestibular cerebellum. Each semicircular canal afferent innervates one of the three canals and encodes information about angular head motion. Otolith afferents innervate either the utricle or saccule and encode information about translational acceleration. Notably, otolith afferents respond to the inertial forces produced by translational motion through the environment or by changes in head orientation relative to gravity. Both canal and otolith afferent fibers are classified on the basis of the regularity of their resting discharge. In general, irregular afferents have larger axons and preferentially transmit information from either the type I hair cells located at the center of the neuroepithelium (known as C-irregulars) or from

(Continued)

Figure 7.8 (*Continued*)

both type I and type II hair cells (known as dimorphic or D-irregulars), whereas regular afferents preferentially provide bouton (B for the regular afferents) endings to type II hair cells. (B) The vestibular system makes essential contributions to our perception of self-motion and ability to navigate as well as to vital reflex pathways (the vestibulo-ocular reflex (VOR) and the vestibulo-spinal reflex (VSR)). Vestibular information is sent to the cortex via two ascending vestibular thalamocortical pathways: the anterior vestibulo-thalamic pathway, comprised of projections from the vestibular nuclei (VN) to the nucleus prepositus and supragenual nucleus (NPH/SGN) and then on to the anterior dorsal thalamus (ADN) via the head direction (HD) network and the posterior vestibulo-thalamic pathway, comprised of projections from the VN through the ventral posterior lateral nucleus (VPL).

Credit: Fig. 1 of ref. 62.

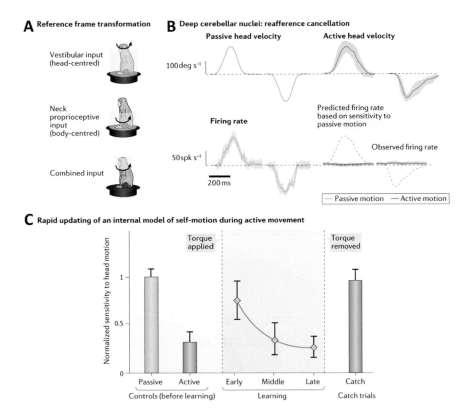

Figure 7.9 Internal models of self-motion in the vestibular cerebellum. (A) Scheme demonstrating stimulation of the vestibular (head-centered) and proprioceptive (body-centered) systems in a monkey. The vestibular
(*Continued*)

Figure 7.9 (*Continued*)

system alone is stimulated by passively rotating the head and body together relative to space (whole-body rotation; top panel) and the proprioceptive system alone is stimulated by passively rotating the body under the head, which remains stationary (middle panel). By contrast, passive rotation of the head relative to the body (bottom panel) produces combined stimulation of the proprioceptive and vestibular systems, thereby allowing us to investigate the transformation of vestibular input from head-centered to body-centered coordinates. (B) The responses of an example deep cerebellar nuclei neuron during passive and active (voluntary) motion paradigms. The top traces illustrate head velocity, the bottom traces show neuronal firing rate responses and the dashed red line indicates a prediction of the firing rate in the active condition based on the neuron's sensitivity to passive motion. The neurons show robust modulation by passive motion, but their responses are minimal when the same motion is actively generated. In the lower panels, gray shading corresponds to the average firing rates and standard deviations for the same 10 movements[63]. Overlaying blue and red lines show the average firing rate responses to passive versus active motion, respectively[63]. (C) When the relationship between the head motor command and resultant movement is altered by applying a resistive load (torque), neuronal vestibular sensitivities to head motion during active head movements initially increase to levels measured during passive head motion. They then gradually decrease to those measured during active head motion before torque application. Once the brain's internal model has been updated to accommodate the new relationship between the voluntary head motor command and the resultant movement, neuronal sensitivities during active trials in which the load is removed (catch trials) are comparable to the neuronal sensitivity during passive head movements[64]. Notably, the re-emergence of afferent suppression during learning, represented by a decrease in the normalized sensitivity of neuronal responses, follows the same time course as the corresponding change in head movement error (not shown)[64].

Credit: Fig. 3 of ref. 62. Part B was adapted by ref. 62 from ref. 63; Part C was adapted by ref. 62 from ref. 64.

7.6 Vector Information from Other Cells Around the Hippocampus

Neurons other than place cells, grid cells and head direction cells (Fig. 7.10) also help define the space, whether egocentric or allocentric (world-centered). There are boundary vector cells devoted for the boundaries both in front of and behind the rat

(Fig. 7.11). Objects in the field are also taken care of by their object vector cells[65].

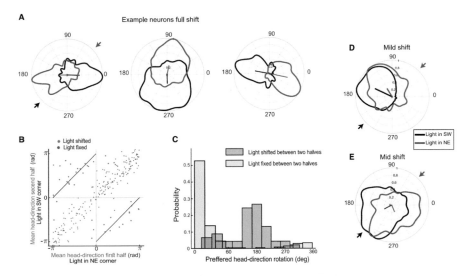

Figure 7.10 Response of hippocampal head direction (HD) cells in quails to cue rotation. Single-unit recordings in the hippocampal formation of Japanese quails (*Coturnix japonica*) were performed with an implanted microdrive, as the quails freely explored a 1 × 1 m open field. (A) Examples of HD tuning curves of three cells before and after 180° cue rotation (light-source shift). Red curves were measured when the light source was in the north-east corner of the arena and black curves when the light source was in the south-west corner (the arrows in the left panel designate the directions of the light sources in the arena). Straight lines originating from the center designate the directions and lengths of the corresponding Rayleigh vectors. (B) Preferred head directions in the second half of the session are plotted versus the preferred head directions in the first half of the session. The blue dots show the results when the light source was shifted at mid-session by 180°. For comparison, superimposed in red are the dots from Figure 3D to show the result without a cue shift. The black diagonal lines indicate the predicted 180° shift of the preferred direction. (C) Distribution of the differences between preferred head directions measured in the two halves of the session, when the light was shifted by 180° (blue), compared to no shift (pink). (D) Example of an HD cell whose tuning did not shift direction with the light-source shift. Plotted as in (A). (E) Example from a different neuron showing an HD tuning curve that shifted about halfway from the predicted shift. Plotted as in (A).

Credit: Fig. 4 of ref. 66.

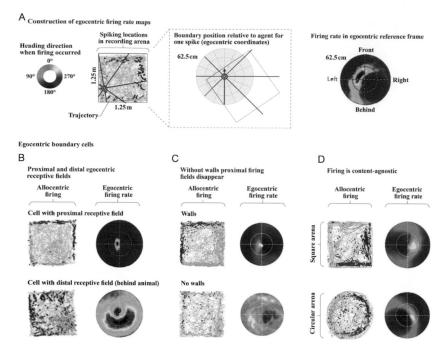

Figure 7.11 Egocentric vector coding cell. Several types of vector coding neurons with egocentric receptive fields have recently been identified. (A) An example of ecogentric coding. The left panel shows the trajectory of an animal as it explores a square arena and the locations at which firing occurred in an egocentric boundary cell in the striatum (dots, color coded according to the animal's heading direction at the time of firing)[67]. The red circle and black arrow show the animal's location and heading direction when one selected spike was fired. The middle panel shows the boundary positions (red lines) relative to the animal when this spike was fired, plotted in an egocentric reference frame in which the animal is considered stationary at the center of the plot. The right panel shows the firing rate map of the cell in egocentric coordinates, created by plotting the spiking frequency in response to the presence of a boundary. A localized maximum in firing rate (a firing field, warm colors) indicates a high firing rate whenever the egocentric coordinates of a boundary coincide with the location of the firing field. (B) Two examples of striatal egocentric boundary cells. In the top panel, the color-coded spikes show that firing was restricted to a given movement direction for each wall (same color coding as in part A). The egocentric firing rate map shows firing for proximate walls on the left of the animal. The bottom panels show another egocentric boundary cell, which has a large egocentric receptive field located behind the animal (sensing the wall behind its back). (C) A retrosplenial egocentric boundary cell with a proximal receptive field loses its firing when the walls of the recording arena are removed. Distal fields persist when walls are removed (not

(Continued)

Figure 7.11 (*Continued*)

shown)[68]. (D) Retrosplenial egocentric boundary cells fire independently of environmental shape[68].

Credit: Fig. 3a-d of ref. 65. Parts A,B were from ref. 67; Parts C,D were from ref. 68.

7.7   Goal-Directed Vector Navigation

Grid cells are oriented in multiple directions and can represent vectors oriented by θ waves (Section 7.1), perhaps with the help of other cells when aligned (Figs. 7.6, 7.10 and 7.11). The firing field of MEC and hippocampal cells shifts to align with goals and rewards[69, 70] (Fig. 7.12). Before running towards an object, an animal measures out the distance from its current position[11, 24]. A study in bat showed that the distance of the chosen path and the path's angle from the goal both modulated hippocampal place cell activity, while the impossible direct distance that had to go through a curtain did not matter[71] (Fig. 7.13). To successfully reach a goal, a stable mental representation of the goal is likely necessary[72] (more in Chapter 9).

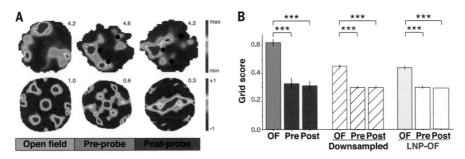

Figure 7.12 Grid score degradation on the cheeseboard. (A) Example of MEC grid cell, which exhibits its highest grid score on the open field and a degraded score on pre- and post-probes. Black dots represent the food rewards. Top: Rate maps with maximum firing rate in red and lowest in blue; peak firing rate (Hz) is shown at upper right. Bottom: Corresponding spatial autocorrelogram maps, ranging from +1 (red) to−1 (blue), with grid score shown at upper right. (B) Average grid score (±SEM) across MEC cells in open field (OF; dark cyan), pre-probe (Pre; blue), and post-probe (Post; red) ($P < 0.00001$, one-way ANOVA). Hatched bars show downsampled control data ($P < 0.00001$, one-way ANOVA). Light cyan and empty bars show control data obtained with LNP spiking model in open field (LNP-OF) ($P < 0.00001$, one-way ANOVA). ***$P <$ 0.001. (one-way ANOVA). Error bars represent SEM. *$P < 0.05$, ***$P < 0.001$.

Credit: Fig. 2 of ref. 69.

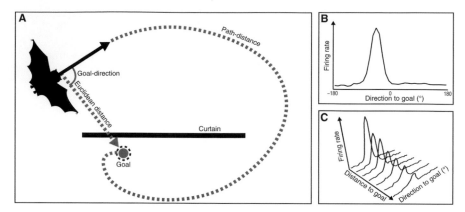

Figure 7.13 Vectorial representation of spatial goals in the hippocampus of bats. (A) Plan view of the experimental set up with bat (not shown to scale) and one of the potential goal locations shown[71]. Three male Egyptian fruit bats each had recordings from 309 cells in the hippocampal area CA1 and flew in a 72 m³ room in search of food. The food platform was placed either in the center of the room or hidden in a randomized location behind an opaque curtain that blocked sight and echolocation (shown here). (B) Illustrative example of a goal-direction neuron with activity plotted against egocentric goal-direction. (C) An illustrative example of a conjunctive distance *x* direction goal-vector cell. 16% of CA1 cells were shown to be modulated by the distance between the goal and the bat's location (goal-distance cells). Many of these cells were more tuned to the path-distance to the goal than the Euclidean distance (A and C). Some cells displayed a conjunctive representation of both goal-direction and goal-distance, thus displaying a vectorial encoding of the goal.

Credit: Fig. 1 of ref. 73.

7.8   A More Versatile Generative Adversarial Network (GAN) in the Brain?

I discussed in Chapter 4 that the classic model for memory consolidation should get a better understanding of the searching and hashing process that is likely what the hippocampal cells are really doing when they interact with the cortex (Fig. 4.11), both when awake and in NREM sleep (Chapter 6); and get down to the physical basis of memory in individual synapses, instead of in some ephemeral waves (Chapter 5). We can complete the model here with the MEC grid cells (Fig. 7.14). Information stored in the cortex, including patterns stored through grids, is retrieved as in a Generative Adversarial Networks (GAN) to serve the current experience, and compared with perception (or outcome) to detect discrepancies (Fig. 7.15). The search for related experiences in the brain, precisely as sharp-wave ripples or broader and slower waves of oscillating neurons (e.g., γ waves possibly by interneurons, Section 6.3; θ waves with a ratio of excitatory and inhibitory neurons[74], Fig. 6.10) that modulate the probability of ripples, is not limited to keywords, places, image, or time. A certain posture can suddenly remind oneself of a past episode. There could be other GANs in the neural system.

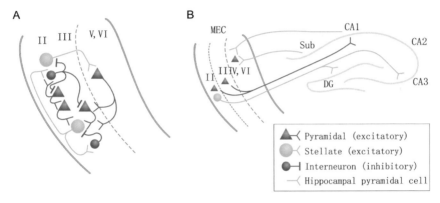

Figure 7.14 Circuits in MEC (A) and between MEC and hippocampus (B). (A) MEC layer II (L2) pyramidal cells make lateral excitatory synapses onto stellate cells, interneurons and other pyramidal cells within L2, whereas L2 stellate cells are thought to communicate with each other primarily via feedforward inhibition through interneurons. (B) MEC stellate cells in L2 project primarily to the dentate gyrus (DG) and CA3, while pyramidal cells in L2 and L3 project to CA1. Reciprocally, hippocampal CA1 and subiculum (sub) pyramidal neurons project primarily back to L5 and L6 of the MEC (A), where grid cells are present but less numerous than in the superficial layers. Stellate cells in MEC L2 project to L5b, while L5a projects to multiple regions of the brain[75]. Location replays in MEC L5/6 were 10 ms delayed compared to replays in hippocampal CA1 neurons[76]. Direct projections from CA1 or the subiculum to L3 and from CA2 to L2 have also been observed, as have minor projections from L5 back to the hippocampus. Ascending excitatory connections from the deep layers to the superficial layers of the MEC close the MEC–hippocampus loop (A).

Credit: Box 1 of ref. 1. CA2 was added for consistency.

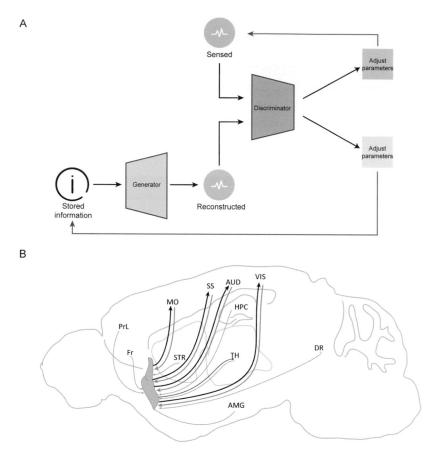

Figure 7.15 A Generative Adversarial Network (GAN) in the brain? (A) Information from the cortex is likely retrieved by the EC-hippocampus to generate the current prediction, and discrepancies are stored, for a better pattern in future runs (see also Chapter 9). Unlike a typical GAN that only feeds back to the generator and the discriminator, the senses can be enhanced for more efficient learning; More stored information could be mobilized (e.g., searched from the cortex with two-dimensional patterns from the MEC), and iteratively updated (e.g., with fine-hashing from the hippocampus, Fig. 4.11), to generate a better reconstruction. Input from the thalamus to the cortex integrates information regarding the current environment[95]. Exposure to novel environments actives the claustrum, which projects to the entire MEC[79], consistent with a role in "attention." The claustrum, continuous with Layer VIc of the insular cortex (Fig. 1.4C), interacts with most regions of the brain[79, 96, 97]. (B) Reciprocal input–output connectivity from the cortex to the claustrum in rodents[96], which could be

(*Continued*)

Figure 7.15 (*Continued*)

part of the discriminator in (A). Additional inputs come from the prefrontal cortex as well as subcortical inputs from the striatum, thalamus, amygdala, and dorsal raphe. Several allocortical (as opposed to isocortical) regions receive inputs from the claustrum, including the piriform cortex, subiculum, and entorhinal cortex, as well as the ventral zones of the prefrontal cortex. The claustrum mostly receives input from Layer VI of the cortex, and mostly sends output to Layer IV pyramidal neuron or interneuron and local interneuron, but also to Layers I/II and VI[96]. Phylogenic differences in the organization of the sensory system of different species mean that homologous areas are not always found. Therefore, the map depicted should be taken as a general outline, consistent with the majority of the data. PrL, prelimbic cortex; ACA, anterior cingulate area; ILA, infralimbic area; MO, motor cortex; SS, somatosensory cortex; ENT, entorhinal area; AI, agranular insular area; PIR, piriform area; PrL, prelimbic area; Fr, frontal cortex; AUD, auditory cortex; VIS, visual cortex; STR, striatum; TH, thalamus; AMG, amygdala; HPC, hippocampus; DR, dorsal raphe.

Credit: Part A, Huijue Jia, Yanzheng Meng. Part B, Fig. 1A of ref. 96.

Hippocampal CA1 θ waves (7 Hz; 2–10 Hz in humans[77]) preceded phase-locked modulation of neocortical γ waves by ~50 ms[78], which could then (perhaps going through the claustrum[79], Figs. 1.4C and 7.15) drive the next round of MEC θ waves[80] (pyramidal cells and inhibitory interneurons), which drives the next round of hippocampal CA3, CA1 θ waves[1, 13, 20, 32, 81, 82] (Figs. 7.14 and 7.6B). PV-expressing interneurons, i.e., basket cells (Fig. 6.6), in the hippocampus are strongly excited by pyramidal cells in superficial layers of the hippocampus that project to the medial prefrontal cortex (mPFC), and these basket cells can then inhibit and synchronize other cells[83]. Inhibitory neurons in the MEC and LEC sparsely synapse on CA1 interneurons, including suppression of CCK-expressing interneurons (Section 6.3) in CA1 that are otherwise excited by CA3 inputs, which gates hippocampal activity and plasticity[84]. Stellate cells in EC L2 (Section 7.2) appeared to fire between CA3 and CA1 θ waves[85]. Besides, MEC layer Va neurons also project to the prefrontal cortex and the basal amygdala[75, 86], which might then reach the hippocampus through the ACC (Section 4.4). During dreaming or day-dreaming (Chapter 6), activity in the cortex (especially the visual cortex) likely goes ahead of that in the hippocampus[87]. Cortical sharp-wave ripples (probably representing synaptic activity, Chapter 5) were recently reported to lead hippocampal sharp-wave ripples during waking hours, while hippocampal ripples lead cortical ripples during NREM sleep (declarative memory formation, e.g., for learning a vocabulary, Fig. 8.14)[88].

After more actual experience or mental replay (preplay), MEC grid cells and hippocampal place cells fired in earlier phases[23, 77, 81, 89, 90], making the animal appear more "prepared" for the place. Hippocampal CA3 input arrives at the trough of CA1 θ waves and drives forward replay (preplay), encouraging new information; The (claustrum-)EC L3 input (Figs. 7.14 and 5.14) arrives at the peak of CA1 θ waves and has been suggested to drive reverse replay (from the current place backward)[91] and supports statistical learning in association with the reward (or pain)[92]. Activity in the mouse primary visual cortex (V1) is also modulated by θ waves, and with experience, can expect things[93, 94].

Unlike a typical GAN that only feeds back to the generator and the discriminator (Fig. 7.15), the senses can be enhanced for more efficient learning. In novel environments, θ frequency increased more slowly with speed[55], and the expanded place cell firing fields likely allowed for finding a match with a lower threshold. More stored information could be mobilized (e.g., from the cortex), and iteratively updated (dendritic computation in hippocampal pyramidal cells that compete for synaptic weights in each replay), to generate a better reconstruction. This more capable GAN likely plays out differently during sleep (e.g., Fig. 6.9) to allow adjustments. Such a grid-patterned searching and memory circuit is likely important for the cognitive functions in Chapters 8, 9.

7.9 Social Navigation

Social navigation also appeared to rely on spatial navigation. Landmarks (e.g., role models) and patterns are also remembered in this form of territory[98]. Mice transplanted with gut microbes from schizophrenia patients showed impaired spatial cognition and social functions (sociability and social novelty)[99, 100], perhaps due to impaired perception combined with malfunctioning grids; and did not stop running as a normal mouse would do, even after half an hour in a new open field[99, 100]. The CA2 region of the hippocampus (between CA1 and CA3, Fig. 7.14), with input from the LEC[101], plays a role in social memory, which is necessary for appropriate behavior[102-105]. Activity in the amygdala was required for knowledge about social rank[106]. Winning or losing a social dominance test in mice was reported to involve activity of VIP-expressing interneurons on PV-expressing interneurons (Figs. 6.5 and 6.6) and thereby pyramidal neurons in the dorsomedial prefrontal cortex (dmPFC)[107]. Downward or upward social comparisons involve the brain regions that have been known for the processing of nonsocial reward or loss, respectively[108]. There would of course be individual differences in the value of rewards and loss.

In a study using a virtual setting, the power of each character and his or her affiliation with oneself was found to be represented together in a 2D space[109]. The study found 2D social navigation to be egocentric instead of allocentric (using a reference frame centered around oneself instead of around others), at least in this virtual setting of trying to get oneself a home and a job in a new town[109]. In a study where the participants chose a business partner for another character, the popularity and competence of the candidates were also represented in 2D (Fig. 7.16), with decodable fMRI (functional Magnetic Resonance Imaging) signals from the EC and hippocampus, which likely engage other regions of the cortex[110]. The reaction time was faster with a larger vector (growth potential (GP) area in Fig. 7.16), e.g., choosing a very capable partner for an underdog[110].

Bats were shown to devote some place cells for allocentric representation of another bat that the observer bat learned from in flight trajectory, while also tracking its own place[111]. Rats who had watched another rat getting a reward in a maze, with some forward and reverse replays in the hippocampal CA1 (Sections 5.4 and 5.5), could then go to the reward site themselves[112]. If some of the place cells are more allocentric than grid cells, are people who remember streets according to shops instead of maps more interested in checking on how other people are doing?

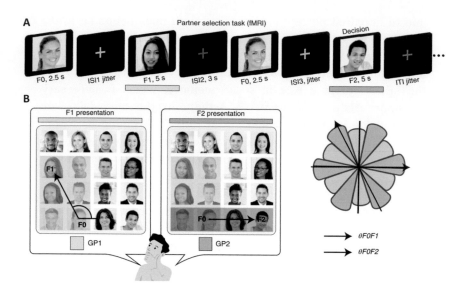

Figure 7.16 Spatial navigation skills used in partner selection task. (A) Illustration of a trial of the partner selection task during fMRI. The participants had learned through feedback-based binary comparisons on the relative rank between pairs of individuals that differ by one rank, each dimension on a separate day, and had to infer the two-dimensional map for the trial task. Participants were then asked to make a binary decision by choosing a better business partner for a given individual (F0) between two (F1 and F2). The better partner is determined by the growth potential (GP) that each pair could expect from their collaboration. Participants could compute the GP of the F0 and F1 pair when F1 is shown and the GP of the F0 and F2 pair when F2 is shown. Participants were subsequently asked to make a decision during F2 presentation. No feedback was given. To compute the GP of a pair, participants were instructed that the GP corresponds to the higher "rank of the pair" in each dimension. Participants were further asked to weigh the "rank of the pair" in both dimensions equally. Therefore, the GP corresponds to the area drawn by the higher rank of the two people in each dimension in the 2D hierarchy (GP_{F0F1} (green rectangle) > GP_{F0F2} (red rectangle); F1 is the better partner for F0 in this example). The authors of the study hypothesized that people would infer direct trajectories over the mentally reconstructed 2D space between the positions of F0 and each potential partner, F1 and F2, to compute the GP for each collaboration searched for neural evidence for hexadirectional modulation of inferred trajectories through the reconstructed cognitive map (θF0F1 at the time of F1 presentation and θF0F2 at the time of F2 presentation)[110]. ISI, inter-stimulus interval; ItI, inter-trial interval.

Credit: Fig. 2 of ref. 110.

7.10   Summary

This Chapter further develops the feedback hashing model for searching and memorizing information introduced in Chapter 4, into a more capable GAN than existing algorithms (Fig. 7.15). Hippocampal place cells hash into the neocortex for detailed information of places (hippocampal CA1 cells, activated by DG-CA3 to store new details, Section 5.5), people (CA2), emotions (amygdala to the anterior hippocampus), concepts and rules (Chapters 8, 9). Sleep (Chapter 6) is perhaps a suitable period to slightly move the hashes to accommodate the ~700 new hippocampal cells every day[113]. The temporal resolution is lower for more distant episodes, which can still be replayed with cortical cues or searched from related stronger tags.

   Grid cells in the medial entorhinal cortex (MEC), on the other hand, work together with the hippocampal cells (increasingly larger place fields along the posterior-anterior axis) to project hexagonal grids, which can accurately lead an animal to new locations. In addition to the 60° highs and 30° lows patterned by grid cells and anchored by place cells, other cells such as head direction cells, boundary vector cells, etc., could help define the orientation and the path. Speed cells might modulate the representation of time. Together, an animal is not bound to a familiar place, and can utilize past experience to explore new places, which are the essence of a GAN (Fig. 7.15). In addition, social navigation takes place in two-dimensional grids and vectors as well. We'll see more abstract and generative use of this navigation system in Chapters 8, 9 for higher level cognition.

Questions

1. How shall we model the appearance of grid-like firing patterns? What types of landmarks and internal states should modulate the firing pattern?

2. How would the different grid sizes impact performance? How often do you think we shift between a small and a large grid when working on a problem? Computers are not limited by gradients in development, so what could be other useful combinations of grids?

3. If a previously learned pattern contains multiple triangles, with some edges having a much higher probability than other edges (some unlikely, some uncertain), based on what we know from Chapters 4–7, how do you think cells in the hippocampus are involved in completing or changing the stored pattern? (Also for Chapter 9)

4. How many cells do you think we need to remember a room? How many of the cells are reused for similar rooms?

5. Can we be more cool-headed about comparing with others, now that we know social navigation is likely also a form of spatial navigation?

References

1. Rueckemann, J. W., Sosa, M., Giocomo, L. M. & Buffalo, E. A. The grid code for ordered experience. *Nat. Rev. Neurosci.* **22**, 637–649 (2021).

2. Hardcastle, K., Maheswaranathan, N., Ganguli, S. & Giocomo, L. M. A multiplexed, heterogeneous, and adaptive code for navigation in medial entorhinal cortex. *Neuron* **94**, 375-387.e7 (2017).

3. Pilly, P. K. & Grossberg, S. Spiking neurons in a hierarchical self-organizing map model can learn to develop spatial and temporal properties of entorhinal grid cells and hippocampal place cells. *PLoS One* **8**, e60599 (2013).

4. Henstridge, C. M., Hyman, B. T. & Spires-Jones, T. L. Beyond the neuron–cellular interactions early in Alzheimer disease pathogenesis. *Nat. Rev. Neurosci.* **20**, 94–108 (2019).

5. L, K. *et al.* Reduced grid-cell-like representations in adults at genetic risk for Alzheimer's disease. *Science* **350**, 430–433 (2015).

6. Lu, K. *et al.* Dissociable effects of APOE ε4 and β-amyloid pathology on visual working memory. *Nat. Aging* (2021) doi:10.1038/s43587-021-00117-4.

7. Ying, J. *et al.* Disruption of the grid cell network in a mouse model of early Alzheimer's disease. *Nat. Commun.* **13**, 886 (2022).

8. Yu, M. *et al.* Structural brain measures linked to clinical phenotypes in major depression replicate across clinical centres. *Mol. Psychiatry* **26**, 2764–2775 (2021).

9. Bellmund, J. L., Deuker, L., Navarro Schröder, T. & Doeller, C. F. Grid-cell representations in mental simulation. *Elife* **5**, e17089 (2016).

10. Navarro Schröder, T., Haak, K. V, Zaragoza Jimenez, N. I., Beckmann, C. F. & Doeller, C. F. Functional topography of the human entorhinal cortex. *Elife* **4**, e06738 (2015).

11. Moser, E. I., Kropff, E. & Moser, M.-B. Place cells, grid cells, and the brain's spatial representation system. *Annu. Rev. Neurosci.* **31**, 69–89 (2008).

12. Hafting, T., Fyhn, M., Molden, S., Moser, M.-B. & Moser, E. I. Microstructure of a spatial map in the entorhinal cortex. *Nature* **436**, 801–6 (2005).

13. Ray, S. *et al.* Grid-layout and theta-modulation of layer 2 pyramidal neurons in medial entorhinal cortex. *Science* **343**, 891–896 (2014).

14. Stensola, T., Stensola, H., Moser, M.-B. & Moser, E. I. Shearing-induced asymmetry in entorhinal grid cells. *Nature* **518**, 207–12 (2015).

15. Rowland, D. C., Roudi, Y., Moser, M.-B. & Moser, E. I. Ten years of grid cells. *Annu. Rev. Neurosci.* **39**, 19–40 (2016).

16. Guanella, A. & Verschure, P. F. M. J. Prediction of the position of an animal based on populations of grid and place cells: a comparative simulation study. *J. Integr. Neurosci.* **6**, 433–46 (2007).

17. Stensola, H *et al.* The entorhinal grid map is discretized. *Nature* **492**, 72–78 (2012).

18. Donato, F., Jacobsen, R. I., Moser, M.-B. & Moser, E. I. Stellate cells drive maturation of the entorhinal-hippocampal circuit. *Science (80-.).* **355**, eaai8178 (2017).

19. Gardner, R. J. *et al.* Toroidal topology of population activity in grid cells. *Nature* **602**, 123–128 (2022).

20. Wang, T. *et al.* Modularization of grid cells constrained by the pyramidal patch lattice. *iScience* **24**, 102301 (2021).

21. Stemmler, M., Mathis, A. & Herz, A. V. M. Connecting multiple spatial scales to decode the population activity of grid cells. *Sci. Adv.* **1**, e1500816 (2015).

22. Zutshi, I., Leutgeb, J. K. & Leutgeb, S. Theta sequences of grid cell populations can provide a movement-direction signal. *Curr. Opin. Behav. Sci.* **17**, 147–154 (2017).

23. Bush, D. & Burgess, N. Advantages and detection of phase coding in the absence of rhythmicity. *Hippocampus* **30**, 745–762 (2020).

24. Bush, D., Barry, C., Manson, D. & Burgess, N. Using grid cells for navigation. *Neuron* **87**, 507–20 (2015).

25. Guanella, A., Kiper, D. & Verschure, P. A model of grid cells based on a twisted torus topology. *Int. J. Neural Syst.* **17**, 231–40 (2007).

26. Krupic, J., Burgess, N. & O'Keefe, J. Neural representations of location composed of spatially periodic bands. *Science (80-.).* **337**, 853–857 (2012).

27. Whittington, J. C. R. *et al.* The Tolman-Eichenbaum machine: unifying space and relational memory through generalization in the hippocampal formation. *Cell* **183**, 1249–1263.e23 (2020).

28. Ginosar, G. *et al.* Locally ordered representation of 3D space in the entorhinal cortex. *Nature* **596**, 404–409 (2021).

29. Gong, Z. & Yu, F. A plane-dependent model of 3D grid cells for representing both 2D and 3D spaces under various navigation modes. *Front. Comput. Neurosci.* **15**, 739515 (2021).

30. Goldshtein, A., Harten, L. & Yovel, Y. Mother bats facilitate pup navigation learning. *Curr. Biol.* **32**, 350–360.e4 (2022).

31. Rodríguez, F. *et al.* Spatial cognition in teleost fish: strategies and mechanisms. *Animals* **11**, 2271 (2021).

32. Gu, Y. *et al.* A map-like micro-organization of grid cells in the medial entorhinal cortex. *Cell* **175**, 736-750.e30 (2018).

33. Naumann, R. K. *et al.* Conserved size and periodicity of pyramidal patches in layer 2 of medial/caudal entorhinal cortex. *J. Comp. Neurol.* **524**, 783–806 (2016).

34. Markram, H. *et al.* Interneurons of the neocortical inhibitory system. *Nat. Rev. Neurosci.* **5**, 793–807 (2004).

35. Tang, Q. *et al.* Pyramidal and stellate cell specificity of grid and border representations in layer 2 of medial entorhinal cortex. *Neuron* **84**, 1191–1197 (2014).

36. Krupic, J., Bauza, M., Burton, S., Barry, C. & O'Keefe, J. Grid cell symmetry is shaped by environmental geometry. *Nature* **518**, 232–235 (2015).

37. Diehl, G. W., Hon, O. J., Leutgeb, S. & Leutgeb, J. K. Grid and nongrid cells in medial entorhinal cortex represent spatial location and environmental features with complementary coding schemes. *Neuron* **94**, 83-92.e6 (2017).

38. Winterer, J. *et al.* Excitatory microcircuits within superficial layers of the medial entorhinal cortex. *Cell Rep.* **19**, 1110–1116 (2017).

39. Chaudhuri, R., Gerçek, B., Pandey, B., Peyrache, A. & Fiete, I. The intrinsic attractor manifold and population dynamics of a canonical cognitive circuit across waking and sleep. *Nat. Neurosci.* **22**, 1512–1520 (2019).

40. Rybakken, E., Baas, N. & Dunn, B. Decoding of neural data using cohomological feature extraction. *Neural Comput.* **31**, 68–93 (2019).

41. Miao, C., Cao, Q., Moser, M.-B. & Moser, E. I. Parvalbumin and somatostatin interneurons control different space-coding networks in the medial entorhinal cortex. *Cell* **171**, 507–521.e17 (2017).

42. Dannenberg, H., Lazaro, H., Nambiar, P., Hoyland, A. & Hasselmo, M. E. Effects of visual inputs on neural dynamics for coding of location and running speed in medial entorhinal cortex. *Elife* **9**, 1–34 (2020).

43. Høydal, Ø. A., Skytøen, E. R., Andersson, S. O., Moser, M.-B. & Moser, E. I. Object-vector coding in the medial entorhinal cortex. *Nature* **568**, 400–404 (2019).

44. Andersson, S. O., Moser, E. I. & Moser, M.-B. Visual stimulus features that elicit activity in object-vector cells. *Commun. Biol.* **4**, 1219 (2021).

45. Campbell, M. G., Attinger, A., Ocko, S. A., Ganguli, S. & Giocomo, L. M. Distance-tuned neurons drive specialized path integration calculations in medial entorhinal cortex. *Cell Rep.* **36**, 109669 (2021).

46. Beed, P. *et al.* Analysis of excitatory microcircuitry in the medial entorhinal cortex reveals cell-type-specific differences. *Neuron* **68**, 1059–1066 (2010).

47. MacDonald, C. J., Lepage, K. Q., Eden, U. T. & Eichenbaum, H. Hippocampal 'time cells' bridge the gap in memory for discontiguous events. *Neuron* **71**, 737–749 (2011).

48. Kraus, B. J., Robinson, R. J., White, J. A., Eichenbaum, H. & Hasselmo, M. E. Hippocampal 'time cells': time versus path integration. *Neuron* **78**, 1090–101 (2013).

49. Cohn-Sheehy, B. I. *et al.* The hippocampus constructs narrative memories across distant events. *Curr. Biol.* (2021) doi:10.1016/j.cub.2021.09.013.

50. Tsao, A. *et al.* Integrating time from experience in the lateral entorhinal cortex. *Nature* **561**, 57–62 (2018).

51. Bellmund, J. L., Deuker, L. & Doeller, C. F. Mapping sequence structure in the human lateral entorhinal cortex. *Elife* **8**, e45333 (2019).

52. Strange, B. A., Witter, M. P., Lein, E. S. & Moser, E. I. Functional organization of the hippocampal longitudinal axis. *Nat. Rev. Neurosci.* **15**, 655–69 (2014).

53. Nielson, D. M., Smith, T. A., Sreekumar, V., Dennis, S. & Sederberg, P. B. Human hippocampus represents space and time during

retrieval of real-world memories. *Proc. Natl. Acad. Sci. U. S. A.* **112**, 11078–83 (2015).

54. Poppenk, J., Evensmoen, H. R., Moscovitch, M. & Nadel, L. Long-axis specialization of the human hippocampus. *Trends Cogn. Sci.* **17**, 230–40 (2013).

55. Wells, C. E. *et al.* Novelty and anxiolytic drugs dissociate two components of hippocampal theta in behaving rats. *J. Neurosci.* **33**, 8650–67 (2013).

56. Carvalho, M. M. *et al.* A brainstem locomotor circuit drives the activity of speed cells in the medial entorhinal cortex. *Cell Rep.* **32**, 108123 (2020).

57. Huang, Z. J. & Paul, A. The diversity of GABAergic neurons and neural communication elements. *Nat. Rev. Neurosci.* **20**, 563–572 (2019).

58. Yang, Y. *et al.* Theta-gamma coupling emerges from spatially heterogeneous cholinergic neuromodulation. *PLoS Comput. Biol.* **17**, e1009235 (2021).

59. Kropff, E., Carmichael, J. E., Moser, E. I. & Moser, M.-B. Frequency of theta rhythm is controlled by acceleration, but not speed, in running rats. *Neuron* **109**, 1029–1039.e8 (2021).

60. da Silva, D. E. *et al.* High-intensity interval training in patients with type 2 diabetes mellitus: a systematic review. *Curr. Atheroscler. Rep.* **21**, 8 (2019).

61. Jensen, C. S. *et al.* Exercise as a potential modulator of inflammation in patients with Alzheimer's disease measured in cerebrospinal fluid and plasma. *Exp. Gerontol.* **121**, 91–98 (2019).

62. Cullen, K. E. Vestibular processing during natural self-motion: implications for perception and action. *Nat. Rev. Neurosci.* **20**, 346–363 (2019).

63. Brooks, J. X. & Cullen, K. E. The primate cerebellum selectively encodes unexpected self-motion. *Curr. Biol.* **23**, 947–955 (2013).

64. Brooks, J. X., Carriot, J. & Cullen, K. E. Learning to expect the unexpected: rapid updating in primate cerebellum during voluntary self-motion. *Nat. Neurosci.* **18**, 1310–1317 (2015).

65. Bicanski, A. & Burgess, N. Neuronal vector coding in spatial cognition. *Nat. Rev. Neurosci.* **21**, 453–470 (2020).

66. Ben-Yishay, E. *et al.* Directional tuning in the hippocampal formation of birds. *Curr. Biol.* **31**, 2592–2602.e4 (2021).

67. Hinman, J. R., Chapman, G. W. & Hasselmo, M. E. Neuronal representation of environmental boundaries in egocentric coordinates. *Nat. Commun.* **10**, 2772 (2019).

68. AS, A. *et al.* Egocentric boundary vector tuning of the retrosplenial cortex. *Sci. Adv.* **6**, eaaz2322 (2020).

69. Boccara, C. N., Nardin, M., Stella, F., O'Neill, J. & Csicsvari, J. The entorhinal cognitive map is attracted to goals. *Science* **363**, 1443–1447 (2019).

70. Butler, W. N., Hardcastle, K. & Giocomo, L. M. Remembered reward locations restructure entorhinal spatial maps. *Science (80-.).* **363**, 1447–1452 (2019).

71. Sarel, A., Finkelstein, A., Las, L., & Ulanovsky, N. Vectorial representation of spatial goals in the hippocampus of bats. *Science* **355**, 176–180 (2017).

72. Hattori, R. & Komiyama, T. Context-dependent persistency as a coding mechanism for robust and widely distributed value coding. *Neuron* **110**, 502–515.e11 (2022).

73. Cothi, W. de & Spiers, H. J. Spatial cognition: goal-vector cells in the bat hippocampus. *Curr. Biol.* **27**, R239–R241 (2017).

74. Cobb, S. R., Buhl, E. H., Halasy, K., Paulsen, O. & Somogyi, P. Synchronization of neuronal activity in hippocampus by individual GABAergic interneurons. *Nature* **378**, 75–78 (1995).

75. Sürmeli, G. *et al.* Molecularly defined circuitry reveals input-output segregation in deep layers of the medial entorhinal cortex. *Neuron* **88**, 1040–1053 (2015).

76. Ólafsdóttir, H. F., Carpenter, F. & Barry, C. Coordinated grid and place cell replay during rest. *Nat. Neurosci.* **19**, 792–4 (2016).

77. Qasim, S. E., Fried, I. & Jacobs, J. Phase precession in the human hippocampus and entorhinal cortex. *Cell* **184**, 3242–3255.e10 (2021).

78. Sirota, A. *et al.* Entrainment of neocortical neurons and gamma oscillations by the hippocampal theta rhythm. *Neuron* **60**, 683–97 (2008).

79. Kitanishi, T. & Matsuo, N. Organization of the claustrum-to-entorhinal cortical connection in mice. *J. Neurosci.* **37**, 269–280 (2017).

80. Chen, D. *et al.* Theta oscillations coordinate grid-like representations between ventromedial prefrontal and entorhinal cortex. *Sci. Adv.* **7**, eabj0200 (2021).

81. Hafting, T., Fyhn, M., Bonnevie, T., Moser, M.-B. & Moser, E. I. Hippocampus-independent phase precession in entorhinal grid cells. *Nature* **453**, 1248–52 (2008).

82. Zutshi, I., Valero, M., Fernández-Ruiz, A. & Buzsáki, G. Extrinsic control and intrinsic computation in the hippocampal CA1 circuit. *Neuron* **110**, 658–673.e5 (2022).

83. Topolnik, L. & Tamboli, S. The role of inhibitory circuits in hippocampal memory processing. *Nat. Rev. Neurosci.* **23**, 476–492 (2022).

84. Basu, J. *et al.* Gating of hippocampal activity, plasticity, and memory by entorhinal cortex long-range inhibition. *Science (80-.).* **351**, aaa5694 (2016).

85. Mizuseki, K., Sirota, A., Pastalkova, E. & Buzsáki, G. Theta oscillations provide temporal windows for local circuit computation in the entorhinal-hippocampal loop. *Neuron* **64**, 267–280 (2009).

86. Kitamura, T. *et al.* Engrams and circuits crucial for systems consolidation of a memory. *Science* **356**, 73–78 (2017).

87. Ji, D. & Wilson, M. A. Coordinated memory replay in the visual cortex and hippocampus during sleep. *Nat. Neurosci.* **10**, 100–107 (2007).

88. Dickey, C. W. *et al.* Widespread ripples synchronize human cortical activity during sleep, waking, and memory recall. *Proc. Natl. Acad. Sci.* **119**, e2107797119 (2022).

89. Jeewajee, A. *et al.* Theta phase precession of grid and place cell firing in open environments. *Philos. Trans. R. Soc. Lond. B. Biol. Sci.* **369**, 20120532 (2013).

90. Bush, D., Ólafsdóttir, H.F, Barry, C. & Burgess, N. B. Ripple band phase precession of place cell firing during replay. *Curr. Biol.* **32**, 64–73.e5 (2022).

91. Wang, M., Foster, D. J. & Pfeiffer, B. E. Alternating sequences of future and past behavior encoded within hippocampal theta oscillations. *Science* **370**, 247–250 (2020).

92. Schapiro, A. C., Turk-Browne, N. B., Botvinick, M. M. & Norman, K. A. Complementary learning systems within the hippocampus: a neural network modelling approach to reconciling episodic memory with statistical learning. *Philos. Trans. R. Soc. B Biol. Sci.* **372**, 20160049 (2017).

93. Fournier, J. *et al.* Mouse visual cortex is modulated by distance traveled and by theta oscillations. *Curr. Biol.* **30**, 3811–3817.e6 (2020).

94. Fiser, A. *et al.* Experience-dependent spatial expectations in mouse visual cortex. *Nat. Neurosci.* **19**, 1658–1664 (2016).

95. Shepherd, G. M. G. & Yamawaki, N. Untangling the cortico-thalamo-cortical loop: cellular pieces of a knotty circuit puzzle. *Nat. Rev. Neurosci.* **22**, 389–406 (2021).

96. Goll, Y., Atlan, G. & Citri, A. Attention: the claustrum. *Trends Neurosci.* **38**, 486–95 (2015).

97. Peng, H. *et al.* Morphological diversity of single neurons in molecularly defined cell types. *Nature* **598**, 174–181 (2021).

98. Wirth, S. *et al.* Territorial blueprint in the hippocampal system. *Trends Cogn. Sci.* **25**, 831–842 (2021).

99. Zhu, F. *et al.* Metagenome-wide association of gut microbiome features for schizophrenia. *Nat. Commun.* **11**, 1612 (2020).

100. Zhu, F. *et al.* Transplantation of microbiota from drug-free patients with schizophrenia causes schizophrenia-like abnormal behaviors and dysregulated kynurenine metabolism in mice. *Mol. Psychiatry* **25**, 2905–2918 (2020).

101. Lopez-Rojas, J., de Solis, C. A., Leroy, F., Kandel, E. R. & Siegelbaum, S. A. A direct lateral entorhinal cortex to hippocampal CA2 circuit conveys social information required for social memory. *Neuron* **110**, 1559–1572.e4 (2022).

102. Banker, S. M., Gu, X., Schiller, D. & Foss-Feig, J. H. Hippocampal contributions to social and cognitive deficits in autism spectrum disorder. *Trends Neurosci.* **44**, 793–807 (2021).

103. Tzakis, N. & Holahan, M. R. Social memory and the role of the hippocampal CA2 region. *Front. Behav. Neurosci.* **13**, 233 (2019).

104. Pimpinella, D. *et al.* Septal cholinergic input to CA2 hippocampal region controls social novelty discrimination via nicotinic receptor-mediated disinhibition. *Elife* **10**, e65580 (2021).

105. Wu, X., Morishita, W., Beier, K. T., Heifets, B. D. & Malenka, R. C. 5-HT modulation of a medial septal circuit tunes social memory stability. *Nature* **599**, 96–101 (2021).

106. Kumaran, D., Melo, H. L. & Duzel, E. The emergence and representation of knowledge about social and nonsocial hierarchies. *Neuron* **76**, 653–66 (2012).

107. Zhang, C. *et al.* Dynamics of a disinhibitory prefrontal microcircuit in controlling social competition. *Neuron* **110**, 516–531.e6 (2022).

108. Luo, Y., Eickhoff, S. B., Hétu, S. & Feng, C. Social comparison in the brain: a coordinate-based meta-analysis of functional brain imaging

studies on the downward and upward comparisons. *Hum. Brain Mapp.* **39**, 440–458 (2018).

109. Tavares, R. M. *et al.* A map for social navigation in the human brain. *Neuron* **87**, 231–43 (2015).

110. Park, S. A., Miller, D. S. & Boorman, E. D. Inferences on a multi-dimensional social hierarchy use a grid-like code. *Nat. Neurosci.* **24**, 1292–1301 (2021).

111. Omer, D. B., Maimon, S. R., Las, L. & Ulanovsky, N. Social place-cells in the bat hippocampus. *Science (80-.).* **359**, 218–224 (2018).

112. Mou, X., Pokhrel, A., Suresh, P. & Ji, D. Observational learning promotes hippocampal remote awake replay toward future reward locations. *Neuron* **110**, 891–902.e7 (2022).

113. Spalding, K. L. *et al.* Dynamics of hippocampal neurogenesis in adult humans. *Cell* **153**, 1219–1227 (2013).

Chapter 8

Arithmetics, Talking and Reading

Abstract

After discussing spatial navigation and social navigation in Chapter 7, we move on to the functions that are historically regarded as human only, arithmetics and language in this chapter. Does the number of different things we can hold in working memory relate to the famous problem of coloring a map with different colors without mixing up? How are numbers, mathematical operations, or letters and words represented in the brain? To talk fluently, the correct patterns and time intervals would need to be prepared and rendered. Despite engaging some different brain regions for particular cognitive or motor functions, the overall answer is potentially coherent from Chapters 7 to 9.

Keywords

Decentralized computation, Blockchain, Object tracking, Working memory, Torus, 7-color theorem, Number neuron, Neuronal tuning, Mutual information, Language

8.1   A Distributed Network That Works Together

Dehaene *et al.* proposed in 1998 a global workspace in the brain[1] (Fig. 8.1). Neurons in the original proposal are organized in a top-down manner, while I tend to believe they are more bottom-

Neuroscience for Artificial Intelligence
Huijue Jia
Copyright © 2023 Jenny Stanford Publishing Pte. Ltd.
ISBN 978-981-4968-78-2 (Hardcover), 978-1-003-41098-0 (eBook)
www.jennystanford.com

up (e.g., the sense of smell in Chapter 2, memory ripples and preplay from Chapter 5; but attention from the claustrum-medial entorhinal cortex (MEC) could serve as a top-down signal, Fig. 7.15). Now with popular knowledge of the blockchain technology (Fig. 8.2)[2], it may be easier to understand such a distributed global workspace (Fig. 8.1). Each module shows its current status (e.g., maintained by interneurons), which could be coordinated and updated by major events. Low-frequency firing is maintained for working memory[3].

Whichever sense is involved, the global coordination likely makes use of the same higher-order brain regions (e.g., visual areas for dreaming, Chapter 6; the piriform cortex for olfactory spatial representation[4]). One may speculate that the size and the dominant module of such a global workspace of neurons could vary among individuals and across physiological states. Working memory, including ordered transition between brain states, is modulated by dopamine in the prefrontal cortex, and is deranged in schizophrenia patients[5].

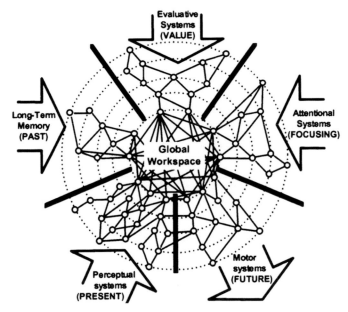

Figure 8.1 The schematic representation of the five main types of processors connected to the global workspace, proposed by ref. 1, inspired by ref. 6.

Credit: Upper part of Fig. 1 in ref. 1. Copyright (1998) National Academy of Sciences, U.S.A.

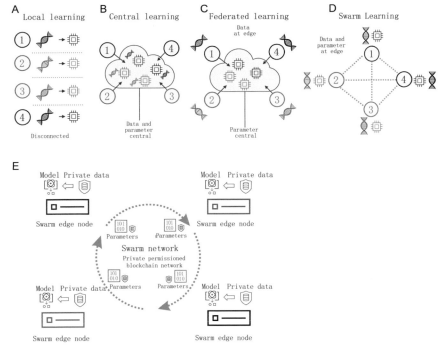

Figure 8.2 Concept of Swarm Learning. (A) Illustration of the concept of local learning with data and computation at different, disconnected locations. (B) Principle of cloud-based machine learning. (C) Federated learning, with data being kept with the data contributor and computing performed at the site of local data storage and availability, but parameter settings orchestrated by a central parameter server. (D) Principle of SL without the need for a central custodian. (E) Schematic of the Swarm network, consisting of Swarm edge nodes that exchange parameters for learning, which is implemented using blockchain technology. Private data are used at each node together with the model provided by the Swarm network.

Credit: Fig. 1a-e of ref. 2.

8.2 Object Tracking for Low Numbers

In many human languages, number three means many. Whether it is the number of syllables in English words (see Section 8.7 regarding the timing), or the number of radicals in a Chinese character, the magic number usually does not exceed three. Studies in human infants suggest that such low numbers are compared as individual objects. E.g., when crackers are presented to infants,

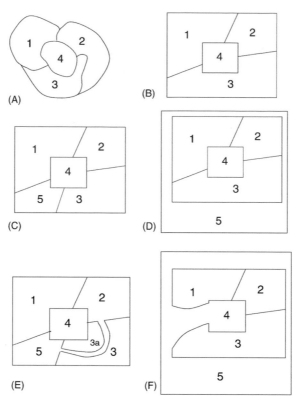

Figure 8.3 Illustration of the four-color principle in plane topology. Four colors (or other markers, indicated here with numeral labels) are sufficient to draw any map, so that adjacent areas are always distinguished along their borders. This principle is adapted here to the consideration of hypothetical patches of cortex that are activated to represent attributes of working memory. It implies that so long as there are no more than four copresent attribute-representations, any one of those representational "subpatches" has free access to make associative contact with any other along a border. (A), (B) Two illustrations of the way in which, with four subareas, each can touch all others along a border. (C), (D) If a fifth subarea is added it is no longer possible for each to contact every other along a border. In (C), subareas 5 and 2 are disconnected. In (D), subareas 5 and 4 are disconnected. (E), (F) Two illustrations of "vain attempts" to circumvent the subpatch-association implication of the four-color rule. In (E), a pseudopod that extends to connect subareas 5 and 2, breaks subarea 3 into disconnected fragments. In (F), a pseudopod that extends to connect subareas 5 and 4 separates regions 1 and 3; if it were positioned a little differently it would break one of the other subareas into disconnected fragments.

Credit: Fig. 1 of ref. 9.

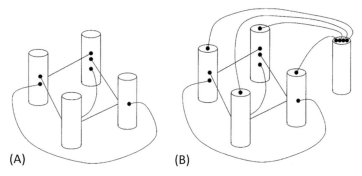

Figure 8.4 Graph-theoretical rules that K_4 is planar and K_5 (a closure of five nodes) is nonplanar illustrated with schematic cortical columns. (A) Every possible connection can be made between four modules without any necessary crossings of connecting lines. There are $C_4^2 = 6$ such connecting lines. (B) When there are five modules, if each is connected with every other, there are $C_5^2 = 10$ such connections. One of these connections must cross over one of the others, no matter how the modules and connecting lines are arranged. If such "overpass connectivity" occurs with the hypothesized cortical working memory representations, the cortex can no longer be considered to be acting strictly as a planar entity.

Credit: Fig. 6 of ref. 9.

the sizes of the crackers also matter[7]. Verification codes sent through cell phones typically have 6 or 4 digits, what percentage of people in the population briefly remembers the code with a single glance (perhaps with silent reading)? Does it take some extra effort to remember the digits in the correct order?

On a planar surface or on a sphere, a minimum of four colors is needed to color each continuous country on a map, known as the 4-color theorem (Figs. 8.3 and 8.4)[8, 9]. And the 4 regions can all be neighboring each other. As soon as there are five modules, they cannot all be on a planar surface and neighboring one another, without cross-passing (Figs. 8.3 and 8.4), i.e., some will always be messed up[8, 9].

8.3 Torus and the Number of Functional Domains on the Hippocampus?

According to graph theory, more colors are needed for the map problem in three dimensions. The donut shape, called a torus,

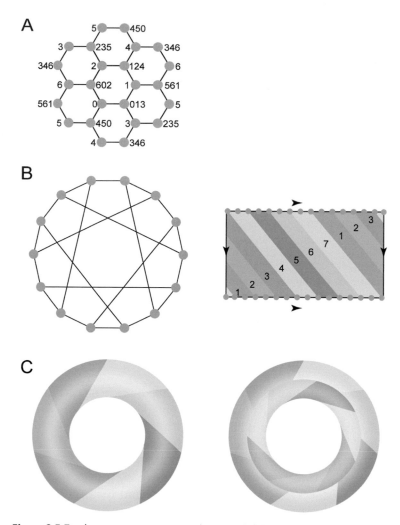

Figure 8.5 7 colors on a torus. A complete graph (all neighboring one another) on seven colors is intrinsically knotted. Yet, on a torus, a minimum of seven colors are needed to color the map. (A) 7 hexagons representing a torus. The Heawood graph in (B) can be drawn onto a torus with no edge crossing. The nodes labeled by three digits show the single-digit nodes (0, 1, 2, 3, 4, 5, 6) they are connected to, repeatedly. (B) Shown by Percy John Heawood in 1890 to prove the 7-color theorem, the Heawood graph (left) is one of eight cubic graphs on 14 nodes with smallest possible graph crossing number of 3. The corresponding 7-color torus map (right) also has 14 nodes on either long edge. (C) Example of a torus with seven colors, which could be derived by folding the rectangle in (B).

Credit: Fei Li.

A

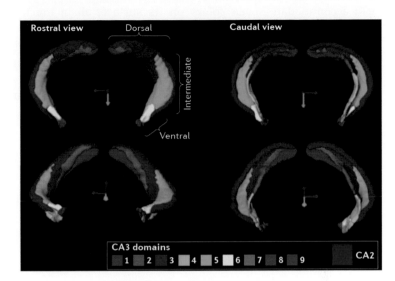

B

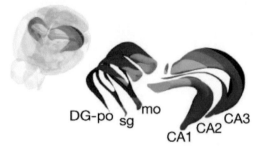

Figure 8.6 Discrete transitions in the molecular, anatomical and functional organization of the hippocampal long axis. (A) Discrete gene expression domains in rat CA3 are defined by reciprocal, nonoverlapping boundaries. Color-coded three-dimensional models of nine gene expression-based subdivisions of CA3 are shown in rostral and caudal views at two different orientations (three-dimensional orientation bars: lateral is red; ventral is green; and rostral is blue). Suggested boundaries for collapsing the nine domains into three domains (ventral, intermediate and dorsal) are indicated in the top left three-dimensional model (compare with Fig. 7.5B). Note, however, that there are substantially different patterns within each of the dorsal, intermediate and ventral domains, and that these are sharp boundaries in some cases. CA2 is indicated in dark blue based on data from 19109908. (B) 3D mouse brain schematic adapted from Allen CCFv3[10] to display the eight dissection regions in hippocampus[11]. Each color represents a dissection region.

Credit: Part A from Fig. 3a of ref. 12; Part B from Extended Fig. 1d of ref. 11.

can accommodate seven colors on its surface, all in touch with one another (Fig. 8.5), which is known as the 7-color theorem for the torus[8]. The left and right hippocampus are twisted planes that look like parts of a torus[9], differing in θ phase in the posterior-anterior axis (dorsal-ventral in rats, Fig. 6.12); The CA3-CA2-CA1 is a continuous extension from the entorhinal cortex (EC), while the dentate gyrus is a separate sheet (Figs. 1.5, 1.6, and 4.9). Interestingly, the hippocampus was reported to have twisted regions on each side (Fig. 8.6, compare with Fig. 8.5). And it looks like the hippocampus became perpendicular, instead of parallel, to the other brain regions only in primates (Section 1.2, Figs. 1.5 and 1.6).

8.4 Analog Representation of Numbers in Humans and Animals

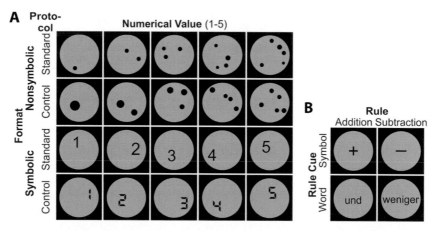

Figure 8.7 Simple addition or subtraction tasks displayed on computer for human participants. After visual fixation on the screen, the first number (operand 1) was followed by a brief delay, after which the addition or subtraction rule was presented, followed in turn by a delay and then the second number (operand 2). After another brief delay, subjects were required to indicate the calculated result (ranging from 0 to 9) on a number panel. (A) Example number stimuli (operand 1 or 2) in the nonsymbolic and symbolic format for standard and control protocols. (B) Example stimuli for the different calculation rules indicated by arithmetic symbols ("+" and "–") and written German words ("und" [add] and "weniger" [subtract]), respectively.

Credit: Fig. 1BC of ref. 19.

Knowing the set size of a group of items (numerosity) underlies everyday decisions of an animal, including decisions to enter combat with a neighboring group[13]. Numbers larger than 4 are typically not tracked as individual objects but are represented by "number cells."

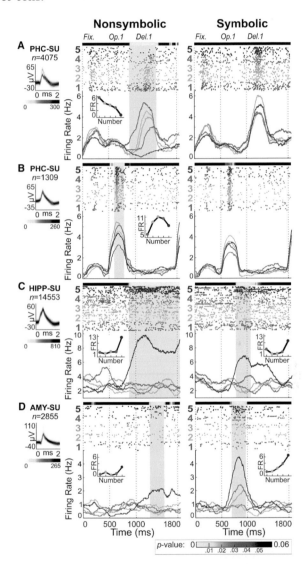

Figure 8.8 Neural responses of number-selective neurons during representation of operand 1 and delay 1, as in the addition or subtraction

(Continued)

Figure 8.8 (*Continued*)

task shown in Fig. 8.7. Nine human subjects undergoing treatment for pharmacologically intractable epilepsy participated in the study, all had bilateral implantation of chronic intracerebral depth electrodes in the medial temporal lobe (MTL) to localize the epileptic focus for possible clinical resection. Across 16 recording sessions from all nine patients, a total of 836 units (585 single and 251 multi units) were identified in the amygdala (AMY; 153 single and 63 multi units), parahippocampal cortex (PHC; 126 single and 61 multi units), entorhinal cortex (EC; 107 single and 54 multi units) and hippocampus (HIPP; 199 single and 73 multi units) according to these criteria; 333 units with firing rates < 1 Hz were excluded. Only single units (representing single neurons) were subjected to further analyses. (A-D) Responses of four example neurons to both nonsymbolic numerosities (left column) and symbolic numerals (right column). Two parahippocampal neurons only responsive to nonsymbolic number with preferred numerosity 1 (A) and 3 (B). A hippocampal neuron (C) and an amygdala neuron (D) responding to both nonsymbolic and symbolic number 5. The left panels depict a density plot of the recorded action potentials (color darkness indicates number of overlapping wave forms according to color scale at the bottom). Panels show single-cell response rasters for many repetitions of the format (each dot represents an action potential) and averaged instantaneous firing rates below. The first 500 ms represent the fixation period. Colors correspond to the five different operand 1 values. Gray shaded areas represent significant number discrimination periods according to the sliding-window ANOVA (color-coded *p*-values above each panel). Insets show the number tuning functions.

Credit: Fig. 2 of ref. 19. Panel D legend corrected from hippocampal to amygdala, after confirmation from the original corresponding author.

Unless trained by some evolutionary pressure, larger numbers are represented with lower precision (Gaussian distribution over a logarithmic numerical scale), as was proposed 100 years ago by Ernst Weber, known as Weber's law[7]. Does that sound familiar? Grid cells and place cells, discussed in Chapter 7 for spatial navigation, is possibly used for numbers. "Number neurons" in and around the hippocampus showed highest activity for a particular number, while also being active for neighboring numbers (Figs. 8.7 and 8.8). The amygdala, which is required for goal-directed behavior and reward consumption (e.g., refs. 14, 15; reinforcement learning in Chapter 9), also contained neurons that can represent numbers and arithmetic operations (Figs. 8.8 and 8.9).

Crows can be trained to include the number zero in their "mental number line", where one-dimensional distances between neurons in the endbrain region nidopallium caudolaterale (NCL) scaled with the numerical differences in countable numbers[16]. Interestingly, the bird NCL is also shaped like a partial donut, in the back of the telencephalon (Fig. 1.1), and contributes to functions such as working memory and serial-order behavior (e.g., refs. 17, 18).

8.5 Abstract Representation of Numbers and Arithmetic Operations

Processing of symbolic number is built upon processing of nonsymbolic number. The symbolic number, in whichever language, would be converted to involve the same cells for mental calculations[19] (Fig. 8.8). While the presented number had the highest peak, the same neuron also responded to neighboring numbers and likely contributed some weights for pinpointing these numbers too (Fig. 8.8). The symbolic number (which would be retrieved by sharp-wave ripples, Section 4.4) appeared to induce a cleaner response curve compared to the nonsymbolic number, and we do not know whether the counting process (e.g., get ready for 6 when counting 5, and the cell for 4 is not off yet), or the visual stimuli including the different sizes and positions of dots contributed to the differences[13, 19].

The arithmetic operations, addition and subtraction, whether shown in symbols or in (German) language, were represented by distinct neurons in the hippocampus, parahippocampal regions, and amygdala[19] (Fig. 8.9). I would not introduce more names by calling them "+ cells" and "- cells." Other rules and concepts are potentially stored in and retrieved from their cells in a similar manner. Many operations and mental efforts can thus become multi-dimensional (chosen on top of 2D, Fig. 8.10).

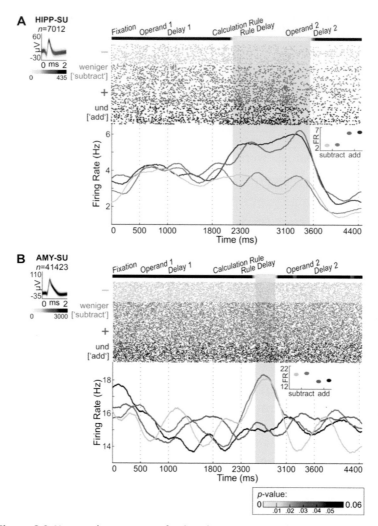

Figure 8.9 Neuronal responses of rule-selective neurons. Responses of two example neurons selective for the "addition" rule (A) and the "subtraction" rule (B). Left panels depict a density plot of the recorded action potentials (color darkness indicates number of overlapping wave forms according to color scale at the bottom). Panels show single-cell response rasters for many repetitions of the rule cue (each dot represents an action potential) and averaged instantaneous firing rates below. Colors correspond to the four different rule cues. Insets show average activity per rule cue during the significant rule discrimination period according to the sliding-window ANOVA (color-coded *p*-values above each panel).

Credit: Fig. S9 of ref. 19.

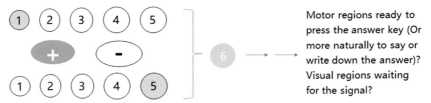

Motor regions ready to press the answer key (Or more naturally to say or write down the answer)? Visual regions waiting for the signal?

Figure 8.10 Schematic diagram for how arithmetic calculations are performed in the brain.

Credit: Huijue Jia.

When we get a result from the mental calculation, the result number needs to be held by some neuron as working memory, while getting ready to push the right button or say the answer. The study did not see activation of the corresponding "number cell" in the MEC grids, which makes sense as the volunteers are not counting[19]. In adults, my best guess would be that the result is held in the form of language, which would involve the auditory or visual projection of that number in the cortex.

A machine learning workflow has recently been developed to help analyze data according to a hypothesis, for mathematicians to develop new theorems[20].

8.6 Multi-Module Coordination during Singing

From insects, birds, to whales and humans, singing has always been a social endeavor. In general, animals use a deeper tone to dominate, and a higher pitch to indicate submissiveness; and imperfections signal emotions[21]. How the rhythm of music compares to a brisk walk or to sleep (or pleasantly in the frequency of delta waves) and modulate the oscillations in the brain (Chapter 6) might deserve some research; at least it is associated with nothing dangerous.

Playing the cello and singing both required brain areas for motor planning and motor execution (Fig. 8.11). The involvement of the motor and premotor cortex also extends to listening to music, which help maintain the rhythm, ready to sing along or tapping[22, 23], perhaps as we replay the activities associated with the music[24] (and emotions and rewards in other brain areas), or re-experience a different space and time from the current reality.

When feedback perturbations are introduced to the pitch, the celloists listened carefully and continued to play well, and the singers compensated for the perturbations with their vocal areas (Fig. 8.12). Singing birds also coordinate with one another when singing together (Fig. 8.13). Talking involves the posterior (/caudal) auditory cortex (Figs. 2.6 and 1.8), and the anterior (/rostral) auditory cortex is suppressed during articulation, compared to its activity when listening to speech[25].

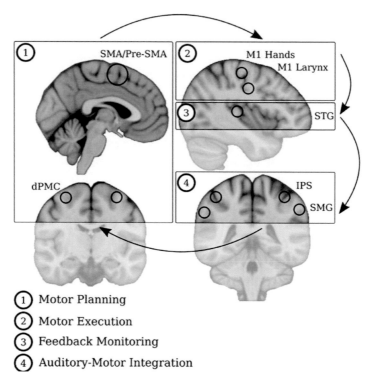

① Motor Planning
② Motor Execution
③ Feedback Monitoring
④ Auditory-Motor Integration

Figure 8.11 Human brain regions active during singing or cello playing. Select components of feedforward and feedback networks. Feedforward regions encompass those related to motor planning (SMA/pre-SMA, (pre-) supplementary motor area; dPMC, dorsal premotor cortex), motor execution (motor cortex M1 Hands, M1 Larynx). Feedback regions encompass those related to feedback monitoring (STG, superior temporal gyrus, including Heschl's gyrus) and auditory-motor integration (IPS, intraparietal sulcus; SMG, supramarginal gyrus).

Credit: Fig. 1 of ref. 26.

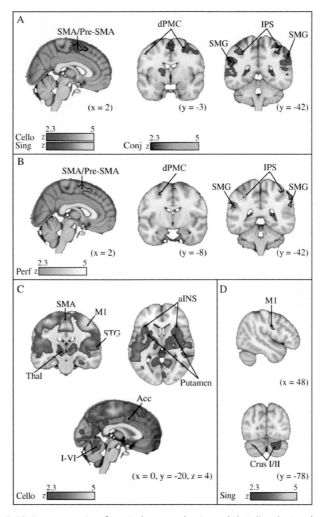

Figure 8.12 Compensating for pitch perturbations. (A) Cello playing (orange), singing (blue), and conjunction (green) of compensate vs simple contrast. (B) Areas within this conjunction (green) where activity positively correlated with good task performance (pink) (C) Cello playing > singing and (D) singing > cello playing. Conjunction and regression showed that SMA/pre-SMA, dPMC, SMG, and IPS all contributed to good task performance in singing and in cello playing; IPS and SMG when compensating for the perturbation, and pSTG and dPMC when ignoring. Cello playing recruited more activity throughout the auditory-motor integration network including in the BG and aINS. Singing shows more activity in vocal areas of motor cortex and cerebellum.

Credit: Fig. 6 of ref. 26.

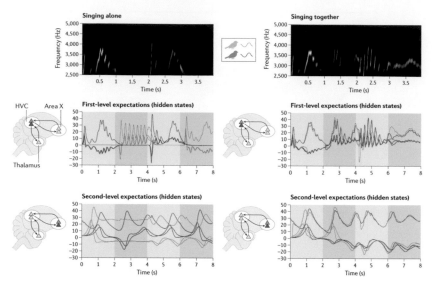

Figure 8.13 Communication using birdsong. Two birds with the same generative models — but different initial conditions (A) — take turns to sing for 2s and then listen for a response (B). The shaded areas indicate which bird is currently singing: red for the first bird and blue for the second bird. When singing, sensory prediction errors are attenuated so that predictions are realized through action. Conversely, when listening, sensory prediction errors are attended by assigning them high precision. The upper panels show the sonogram heard by the first bird (red lines in the lower panels; note that the timescales differ between the upper and the middle/lower graphs). The birds cannot hear each other in (A), while they can in (B). The posterior expectations for the first (red) bird are shown in red as a function of time — and the equivalent expectations for the second (blue) bird are shown in blue. In the left panel, because this bird can hear only itself, the sonogram reflects descending proprioceptive predictions based on expectations in the higher vocal venter (HVC; a premotor region, middle panel) and area X (a higher-order area, lower panel), which projects to the auditory thalamus. The blue and red lines reporting expectations about underlying causes (that is, fluctuations in amplitude and frequency) generating the birdsong are shown for the HVC and area X in the middle and lower panels, respectively. Note that when the birds are listening, their expectations at the first level fall to zero, because they do not hear anything. However, the slower dynamics in area X can generate the song again after the end of each listening period. In the right panel, the two birds can hear each other. In this instance, the expectations show synchrony at both the sensory and the extrasensory hierarchical levels. Note that the sonogram is continuous over successive 2-s epochs — generated alternately by the first bird and the second bird. The key role of precision emerges again; here, in selectively attending to sensory streams — generated by the birds — in a coordinated way that enables turn taking and communication[27, 28].

Credit: From Box 2 of ref. 23, adapted by ref. 23 from ref. 29.

8.7   Talking or Reading

Speech generation by the extended Broca's area (Fig. 1.8) is likely more of a one-dimensional navigation (Fig. 7.3) that is coordinated with breathing and many muscles[21], but often had to choose between alternative paths in pronunciation, wording, and grammar[30]. From baby cries to speech, the vocalization is rhythmic and coordinated within breaths[31]. Syllables occur at a rate of 4–8 Hz, in the same range as hippocampal θ frequencies[27, 32].

Unlike the AI networks for natural language interpretation that are having more and more layers to capture long-range relationships, understanding of spoken or written language by the Wernicke's area (Fig. 1.8) would presumably also use the flat grids that have been learned in the MEC-hippocampus (Fig. 7.15).

We take one second or longer to plan what to say, and jump into conversations with a gap of ~ 0.2 s[25, 33, 34]. By pure speculation, singing and talking would also require the spatial-temporal perception in the MEC-hippocampus, and together with information from the cortex including motor coordination, to generate the working memory plan for the next few words (compared to Generative Adversarial Networks (GAN) in Section 7.8).

On the other hand, reading is likely based on gazing that shifts according to 2D grids[35-37] (Section 8.7.2), (inner) speech interpretation by the extended Wernicke's area (Fig. 1.8) and imagination whenever some keyword matches (Chapter 9).

8.7.1   Hippocampus-Dependent Procedural Memory for Speaking

Studies on individuals learning a second language showed that sleep spindles (NREM sleep, Chapter 6) indicate declarative memory formation, whereas REM sleep is a hallmark of procedural memory consolidation[38] (Fig. 8.14). Just as my high school oral English teacher Julia said, one is truly fluent in English when one starts to dream in it. After memorizing all the bits and pieces[30], the (EC-)hippocampal θ waves and precise synaptic activity seen as sharp-wave ripples (Chapter 5, Fig. 6.12) likely play out a natural flow of language (Fig. 8.14), with learned longer-range patterns reminiscent of birds singing (Fig. 8.13), and coordinated with breathing.

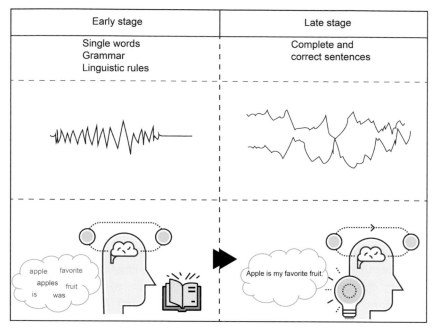

Early stage	Late stage
Single words Grammar Linguistic rules	Complete and correct sentences

Figure 8.14 Role of NREM sleep in declarative memory (left), and role of REM sleep in procedure memory for learning a second language (right).

Credit: Yanzheng Meng.

At least for learning a second language, declarative memory likely includes how new words relate to words in one's mother tongue, how the new words relate to the objects and people they represent, and how the words are correctly pronounced with motor coordination. In toddlers, left hippocampal volume correlates with vocabulary; Left anterior medial temporal lobe (MTL—including the hippocampus on top, entorhinal and perirhinal cortices in the anterior, and the parahippocampal cortex in the posterior) activity, including activation when hearing newly learned words at sleep, indicates a growing vocabulary[39].

Auditory word segmentation relies on transition probability (to choose between alternative paths), which dogs can also been trained to perform[40]. Some of the grammar rules may be not unlike arithmetic operations (Fig. 8.10). Words and grammar from cortical neurons are smoothly combined into sentences with grid-patterned Generative Adversarial Networks (GAN, Section 7.8), and play out often enough so as not to be outcompeted by other dendritic spines (Chapters 5, 6).

8.7.2   Reading from Grids to Details?

There is a Chinese idiom that says "Ten lines at a glance." With enough prior knowledge, the shifts in gazing position observed in monkeys looking at pictures[35-37] (Fig. 8.15; changing orientation and step size, similar to Fig. 9.3) should allow people to grasp the meanings of a paragraph, and look into more details where necessary. Monkeys can be trained to sequentially look at two or three spots of a hexagonal grid in a remembered sequence that was randomly chosen from the six spots each time[41].

Besides established grammatical patterns in a language, the way we tell stories also help readers find and create the proper hash (Fig. 4.11), e.g., people or object, time and space, for storage of new information (also in Chapter 9). For example, a news article in *Nature* begins with "A fossilized tooth unearthed in a cave in northern Laos might have belonged to a young Denisovan girl that died between 164,000 and 131,000 years ago. If confirmed, it would be the first fossil evidence that Denisovans — an extinct hominin species that co-existed with Neanderthals and modern humans — lived in southeast Asia."[42]. The mentioning of Neanderthals and modern humans help group the lesser-known hominin species Denisovans among them, for those who haven't previously saved information on Denisovans.

Cells in the posterior hippocampus have smaller place fields than cells in the anterior hippocampus (one of potential differences among individuals and between animals), and are also closer to the visual cortex (good for economic wiring[43, 44]).

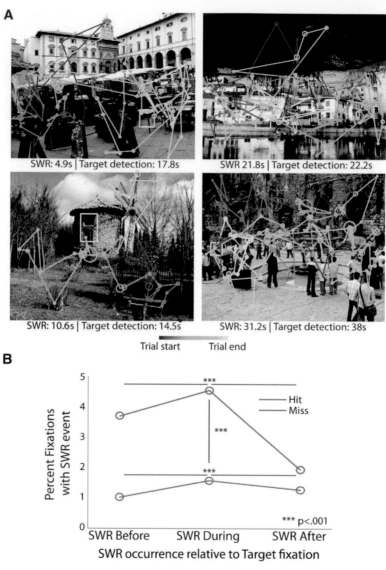

A

SWR: 4.9s | Target detection: 17.8s

SWR 21.8s | Target detection: 22.2s

SWR: 10.6s | Target detection: 14.5s

SWR: 31.2s | Target detection: 38s

Trial start   Trial end

B

Percent Fixations with SWR event

*** Hit
Miss

*** p<.001

SWR Before   SWR During   SWR After

SWR occurrence relative to Target fixation

Figure 8.15 Grid-like shifts in attention as monkeys participated in a visual exploration task[36]. Monkeys (*Macaca mulatta*) m1–m3 were implanted chronically with platinum/tungsten multicore tetrodes lowered into the CA3/DG region of hippocampus; m4 was recorded with 125 and 200 μm tungsten microelectrodes lowered daily, with trajectories aimed at CA3/DG and CA1/subiculum. Eye movements were recorded using video-based eye tracking. (A) Four examples of scan paths during search in successful trials

(*Continued*)

Figure 8.15 (*Continued*)
containing hippocampal sharp-wave ripples (SWRs, Section 4.4; Chapter 5). Fixation (circles) and saccades (lines) with graded coloring according to the scan path (blue, trial start; yellow, trial end), with fixation duration represented by circle size. The nearest fixation in time to the SWR event is shown in black outlined in red, with the flanking fixations and saccades indicated by red only. The elapsed search time when the SWR occurred, and total search time until the target was detected, are listed below each image. Fixation durations during SWRs were ~40 ms longer than for non-SWR fixations (37.73 ms, $z_{(2590)}$ = 9.3, p = 2.8 × 10^{-20}), which is within typical search fixation durations and is more consistent with a local than a global search strategy. In addition, saccade amplitudes were smaller (−0.51° visual angle, $z_{(1049)}$ = −3.3, p = 0.0009), also consistent with local search. (B) SWR occurrence, as a function of scan path distance from fixations directed in the target location. Fractional occurrence is shown separately for trials that were successful or unsuccessful (hits, red; misses, blue). SWRs concurrent with target fixations, shown in the middle group, had more hits than misses (vertical black significance bar). For a given detection type (hit/miss), the percentage of nonconcurrent SWRs (nine fixations preceding, and nine subsequent to, the target fixation window) were compared with the percentage concurrent with target fixations (horizontal black significance bars).

Credit: Fig. 5 of ref. 36.

Recognition of letters or characters can presumably use the same 2D area as face recognition, with some strokes rounder and others pointier (Chapter 2, Fig. 2.10). For example, traditional Chinese characters would be much more difficult to recognize than Latin-based alphabetical languages such as English[45]. The visual decoding system would need to be trained for each language, possibly with tuning of the visual neurons (Fig. 8.16). Neurons in layers 2/3 of the ferret visual cortex receive a varying mixture of input from one or both eyes[46]. The left and right visual cortex have to communicate with each other to recognize symmetrical letters such as "w"[6], which would likely involve projections through the thalamus[47].

Some letters and some handwriting would be easier to recognize, likely due to quantifiable information in the stimulus space (Fig. 8.16). Diseases such as schizophrenia and Parkinson's disease are known to have abnormalities in the retina that manifest as impaired visual processing[48]. And schizophrenia is genetically linked to lower education attainment[49, 50].

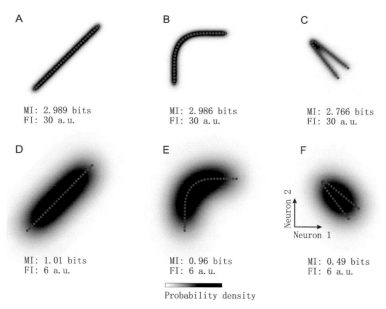

A
MI: 2.989 bits
FI: 30 a.u.

B
MI: 2.986 bits
FI: 30 a.u.

C
MI: 2.766 bits
FI: 30 a.u.

D
MI: 1.01 bits
FI: 6 a.u.

E
MI: 0.96 bits
FI: 6 a.u.

F
MI: 0.49 bits
FI: 6 a.u.

Probability density

Figure 8.16 Relationship between manifold geometry, Fisher information and mutual information. Toy simulation illustrating the encoding of a one-dimensional stimulus space by six different neural codes (A–F). The neural population comprises only two neurons, whose activities define the horizontal and vertical dimensions of each part. For each of 30 equally spaced points in stimulus space, mean response is plotted as a red dot. The neural response manifold is a continuous line (not shown) that passes through the red dots. The response distribution is shown in the background (greyscale) for a uniform prior over the stimulus space. Neural noise (Gaussian and isotropic here) is low in the top row (A–C) and high in the bottom row (D–F). Geometry of the manifold and the noise determine the Fisher information (FI) for each stimulus, the total FI, and the mutual information (MI, defined by entropy) between stimulus and response (see ref. 51 for details). Consider n neurons with tuning functions $fi(\vartheta)$ and independent Gaussian noise, the FI is the squared norm of the gradient of the response pattern with respect to the stimulus variable ϑ. The total FI is proportional to the length of the manifold divided by the standard deviation of the noise. Length of the manifold is the same in all codes (A–F). The total FI is 30 in the low-noise codes (A–C) and 6 in the high-noise codes (parts D–F). Within each noise level (top row, bottom row), the FI is the same whereas the geometry and the MI change. This reflects the fact that the FI measures discriminability along the manifold, whereas geometry and the MI measure discriminability among all stimuli. Folding the manifold (C, F) renders stimuli on opposite banks somewhat confusable, which lowers the MI, but does not affect the FI. The FI, thus, does not determine the geometry or the MI. Numerical values (lower right in each part) show the MI and the total FI.

Credit: Fig. 2 of ref. 51.

An analyses of fMRI data from healthy human subjects working on a matrix-reasoning test (to correctly fill in the missing one, according to a 5×3 matrix) identified brain regions associated with intelligence, including multiple regions that function in language. The authors suggest utilization of inner speech for such reasoning tasks[52]. When reading a book, we probably use these language regions too, with more memory recalls and imagination (more on brain-wide coordination in Chapter 9).

8.8 Summary

Besides the torus of grid cells in Chapter 7, the map color problem for partial torus shapes such as the hippocampus could likely explain the number of different things we can coordinate in working memory. From birds to humans, individual numbers are likely represented along an axis in discrete neurons in and around the hippocampus. Mathematical operations, and perhaps letters, words and grammatical rules are similarly stored in place with ordered patterns and probabilities.

Unlike the AI networks for natural language and image processing that are having more and more layers, flat grids of different sizes can be adaptively incorporated in the brain (more in Chapter 9). The effect (reward or punishment) associated with saying something also matters. Consistent with the MEC grids, hippocampal place cells, and the θ wave/γ wave-driven searching that leads to sharp-wave ripples that competitively modulates synaptic weights (elaborated in Chapters 4–7), declarative memory needs to be smoothly linked as procedural memory for fluent speech.

Relieved from all the muscle coordination and social expectations, reading can be in shifting or consistently sized steps. Tuning of the visual input during development, along with memorized combinations, might help improve the speed of recognition.

There are as yet too many places to record neuronal activities, in model animals or in patients, and much more effort would be needed before we understand how the brain masters such impressive functions as mathematics and language.

Questions

1. What would be the useful axes and patterns (1D and 2D grids, Chapter 7) to remember, when children study mathematics?

2. What would be the useful patterns to remember when children learn the language in your culture? What are the time frames for short-range and long-range interactions that one has to keep in mind?

3. In the big-data era, what are the more natural data that would be useful to study, for topics in this chapter?

References

1. Dehaene, S., Kerszberg, M. & Changeux, J.-P. A neuronal model of a global workspace in effortful cognitive tasks. *Proc. Natl. Acad. Sci.* **95**, 14529–14534 (1998).

2. Warnat-Herresthal, S., *et al.* Swarm Learning for decentralized and confidential clinical machine learning. *Nature* **594**, 265–270 (2021).

3. Kornblith, S., Quiroga, R. Q., Koch, C., Fried, I. & Mormann, F. Persistent single-neuron activity during working memory in the human medial temporal lobe. *Curr. Biol.* **27**, 1026–1032 (2017).

4. Poo, C., Agarwal, G., Bonacchi, N. & Mainen, Z. F. Spatial maps in piriform cortex during olfactory navigation. *Nature* **601**, 595–599 (2022).

5. Braun, U. *et al.* Brain network dynamics during working memory are modulated by dopamine and diminished in schizophrenia. *Nat. Commun.* **12**, 3478 (2021).

6. Mesulam, M. From sensation to cognition. *Brain* **121**, 1013–1052 (1998).

7. Dehaene, S., Molko, N., Cohen, L. & Wilson, A. J. Arithmetic and the brain. *Curr. Opin. Neurobiol.* **14**, 218–24 (2004).

8. White, A. T. *Graphs, Groups, and Surfaces* (Elsevier Science Publishers B.V., 1984).

9. Glassman, R. B. Topology and graph theory applied to cortical anatomy may help explain working memory capacity for three or four simultaneous items. *Brain Res. Bull.* **60**, 25–42 (2003).

10. Wang, Q. *et al.* The Allen Mouse brain common coordinate framework: a 3D reference atlas. *Cell* **181**, 936 (2020).

11. Liu, H. *et al.* DNA methylation atlas of the mouse brain at single-cell resolution. *Nature* **598**, 120–128 (2021).

12. Strange, B. A., Witter, M. P., Lein, E. S. & Moser, E. I. Functional organization of the hippocampal longitudinal axis. *Nat. Rev. Neurosci.* **15**, 655–69 (2014).

13. Nieder, A. Counting on neurons: the neurobiology of numerical competence. *Nat. Rev. Neurosci.* **6**, 177–90 (2005).

14. Janak, P. H. & Tye, K. M. From circuits to behaviour in the amygdala. *Nature* **517**, 284–292 (2015).

15. Courtin, J. *et al.* A neuronal mechanism for motivational control of behavior. *Science (80-.).* **375**, eabg7277 (2022).

16. Kirschhock, M. E., Ditz, H. M. & Nieder, A. Behavioral and neuronal representation of numerosity zero in the crow. *J. Neurosci.* **41**, 4889–4896 (2021).

17. Güntürkün, O. The avian 'prefrontal cortex' and cognition. *Curr. Opin. Neurobiol.* **15**, 686–693 (2005).

18. Johnston, M., Clarkson, A. N., Gowing, E. K., Scarf, D. & Colombo, M. Effects of nidopallium caudolaterale inactivation on serial-order behavior in pigeons (Columba livia). *J. Neurophysiol.* **120**, 1143 1152 (2018).

19. Kutter, E. F., Bostroem, J., Elger, C. E., Mormann, F. & Nieder, A. Single neurons in the human brain encode numbers. *Neuron* **100**, 753–761. e4 (2018).

20. Davies, A. *et al.* Advancing mathematics by guiding human intuition with AI. *Nature* **600**, 70–74 (2021).

21. Colapinto, J. *This Is the Voice* (Simon & Schuster, 2021).

22. Cannon, J. J. & Patel, A. D. How beat perception co-opts motor neurophysiology. *Trends Cogn. Sci.* **25**, 137–150 (2021).

23. Vuust, P., Heggli, O. A., Friston, K. J. & Kringelbach, M. L. Music in the brain. *Nat. Rev. Neurosci.* **23**, 287–305 (2022).

24. Wandelt, S. K. *et al.* Decoding grasp and speech signals from the cortical grasp circuit in a tetraplegic human. *Neuron* **110**, 1777–1787. e3 (2022).

25. Jasmin, K., Lima, C. F. & Scott, S. K. Understanding rostral–caudal auditory cortex contributions to auditory perception. *Nat. Rev. Neurosci.* **20**, 425–434 (2019).

26. Segado, M., Zatorre, R. J. & Penhune, V. B. Effector-independent brain network for auditory-motor integration: fMRI evidence from singing and cello playing. *Neuroimage* **237**, 118128 (2021).

27. Ghazanfar, A. A. & Takahashi, D. Y. The evolution of speech: vision, rhythm, cooperation. *Trends Cogn. Sci.* **18**, 543–53 (2014).

28. Wilson, M. & Wilson, T. P. An oscillator model of the timing of turn-taking. *Psychon. Bull. Rev.* **12**, 957–68 (2005).

29. Friston, K. J. & Frith, C. D. Active inference, communication and hermeneutics. *Cortex.* **68**, 129–143 (2015).

30. Ullman, M. T. A neurocognitive perspective on language: the declarative/ procedural model. *Nat. Rev. Neurosci.* **2**, 717–726 (2001).

31. Wei, X. P., Collie, M., Dempsey, B., Fortin, G. & Yackle, K. A novel reticular node in the brainstem synchronizes neonatal mouse crying with breathing. *Neuron* **110**, 644–657.e6 (2022).

32. Guilleminot, P. & Reichenbach, T. Enhancement of speech-in-noise comprehension through vibrotactile stimulation at the syllabic rate. *Proc. Natl. Acad. Sci.* **119**, e2117000119 (2022).

33. Norman, Y. *et al.* Hippocampal sharp-wave ripples linked to visual episodic recollection in humans. *Science* **365**, eaax1030 (2019).

34. Castellucci, G. A., Kovach, C. K., Howard, M. A., Greenlee, J. D. W. & Long, M. A. A speech planning network for interactive language use. *Nature* **602**, 117–122 (2022).

35. Killian, N. J., Jutras, M. J. & Buffalo, E. A. A map of visual space in the primate entorhinal cortex. *Nature* **491**, 761–764 (2012).

36. Leonard, T. K. *et al.* Sharp wave ripples during visual exploration in the primate hippocampus. *J. Neurosci.* **35**, 14771–14782 (2015).

37. MLR, M. & EA, B. Neurons in primate entorhinal cortex represent gaze position in multiple spatial reference frames. *J. Neurosci.* **38**, 2430–2441 (2018).

38. K, T. *et al.* Sleep and second-language acquisition revisited: the role of sleep spindles and rapid eye movements. *Nat. Sci. Sleep* **13**, 1887–1902 (2021).

39. Johnson, E. G., Mooney, L., Graf Estes, K., Nordahl, C. W. & Ghetti, S. Activation for newly learned words in left medial-temporal lobe during toddlers' sleep is associated with memory for words. *Curr. Biol.* **31**, 5429–5438.e5 (2021).

40. Boros, M. *et al.* Neural processes underlying statistical learning for speech segmentation in dogs. *Curr. Biol.* **31**, 5512–5521.e5 (2021).

41. Xie, Y. *et al.* Geometry of sequence working memory in macaque prefrontal cortex. *Science (80-.).* **375**, 632–639 (2022).

42. Kreier, F. Ancient tooth suggests Denisovans ventured far beyond Siberia. *Nature* **605**, 602–603 (2022).

43. Bullmore, E. & Sporns, O. The economy of brain network organization. *Nat. Rev. Neurosci.* **13**, 336–349 (2012).

44. Arcaro, M. J. & Livingstone, M. S. On the relationship between maps and domains in inferotemporal cortex. *Nat. Rev. Neurosci.* **22**, 573–583 (2021).

45. Tsu, J. *Kingdom of Characters : the Language Revolution that Made China Modern.* (Riverhead Books, 2022).

46. Scholl, B. *et al.* A binocular synaptic network supports interocular response alignment in visual cortical neurons. *Neuron* **110**, 1573–1584.e4 (2022).

47. Shepherd, G. M. G. & Yamawaki, N. Untangling the cortico-thalamo-cortical loop: cellular pieces of a knotty circuit puzzle. *Nat. Rev. Neurosci.* **22**, 389–406 (2021).

48. Silverstein, S. M. & Rosen, R. Schizophrenia and the eye. *Schizophr. Res. Cogn.* **2**, 46–55 (2015).

49. Trampush, J. W. *et al.* GWAS meta-analysis reveals novel loci and genetic correlates for general cognitive function: a report from the COGENT consortium. *Mol. Psychiatry* **22**, 336–345 (2017).

50. Le Hellard, S. *et al.* Identification of gene loci that overlap between schizophrenia and educational attainment. *Schizophr. Bull.* **43**, 654–664 (2017).

51. Kriegeskorte, N. & Wei, X.-X. Neural tuning and representational geometry. *Nat. Rev. Neurosci.* **22**, 703–718 (2021).

52. Fraenz, C. *et al.* Interindividual differences in matrix reasoning are linked to functional connectivity between brain regions nominated by Parieto-Frontal Integration Theory. *Intelligence* **87**, 101545 (2021).

Chapter 9

Causality and Cognitive Exploration

Abstract

Animals could perform a variety of tasks reasonably well at the first trial. Such adaptive learning likely also depend on the navigation system discussed in the earlier chapters. Rooted in emerging evidence, this chapter proposes strategies that allow the human brain to do what current artificial intelligence programs cannot do. Causal reasoning, including the evaluation of counterfactuals, is intrinsic to the neuronal setup. Useful "common sense" rules would be learned, instead of being crammed in. It might also be rewarding to fill in a missing gap, and accumulate useful patterns for the future, the essence of curiosity.

Keywords

Reinforcement learning, Transfer learning, Cognition, Inferential reasoning, Causal reasoning, Artificial intelligence (AI), Counterfactual, Adaptive learning, Spatial navigation, Generative Adversarial Networks (GAN)

9.1 To Explore or Not

As noted in Chapter 1, to devote energy to the brain is an expensive investment. Not many species have the luxury of thinking about things that are not immediately related to survival[1]; and

Neuroscience for Artificial Intelligence
Huijue Jia
Copyright © 2023 Jenny Stanford Publishing Pte. Ltd.
ISBN 978-981-4968-78-2 (Hardcover), 978-1-003-41098-0 (eBook)
www.jennystanford.com

it helps to mature later[2]. Any two neurons in the neocortex is only two or three synapses away from one another[3], making an ever more impressive framework for all possible associations between events and between concepts, as the brain gets larger (e.g., $10^{10} \times 10^{10}$). Neurons in different layers of the neocortex, together with the thalamus and the claustrum, constitute many parameters that can be adjusted[4, 5].

An intriguing possibility is that weak associations do not involve synapses on dendritic spines[6] (Chapter 5; e.g., parallel axons from one or more interneurons to multiple pyramidal neurons[7], Figs. 6.6 and 6.7), while causality (path diagrams in Section 9.3) is more hard-wired to the point as a series of synapses on dendritic spines. As formation of synapses depends on proximity to available membrane surfaces[3, 8], interneurons might occupy places before these surfaces and space are allocated to pyramidal neurons for more defined links. Interneurons, with their larger synapses on dendritic shafts that release GABA (γ-aminobutyric acid), are important candidates that coordinate localized firing of a group of cortical pyramidal cells[9-12] (Figs. 1.9, 2.14, 6.6, and 6.7). Interneurons are modulated by task (e.g., refs. 13–15), and they participate in large-scale oscillations during sleep that likely regularize the local pattern[16, 17] (Chapter 6). Such synchronizing inhibition might also move dendritic spines along the branch closer to or farther away from the cell body, which would also adjust the electric resistance (Fig. 5.2) and thereby the synaptic weight in neuronal computation. Hippocampal and entorhinal interneurons likely also help determine the assignment of neurons into cues and memories[15, 18, 19] (Chapters 5, 7). The childish large-scale oscillations at a stimulus are trained over the years into discrete small clusters that respond fast (Chapter 3).

Whatever we enjoy doing (e.g., olfactory associative learning facilitated by dopamine, neuronal representation of loss functions[6, 20-23]) must be rewarding in some way, consistent with reinforcement learning (Fig. 9.1; Meta-reinforcement learning[24]). Replay/preplay driven by sharp-wave ripples when awake is modulated by environmental novelty[25, 26] (to more easily get a match with existing experience and to store the differences, Figs. 4.11 and 7.15), and by reward[27, 28] (Figs. 5.12, 7.12, and 9.1). The rates of reverse replay (starting from the target place and back) reflected relative reward magnitudes[27]. Consistently, reverse replay has been suggested to be driven by monosynaptic input from the entorhinal cortex that arrives at the peak of hippocampal CA1 θ waves[29], which updates the synaptic weight[30] (Section 5.4, 5.5). The apparent result of such reinforcement learning could be a more feedforward and modular network (Fig. 3.5) that fires in anticipation.

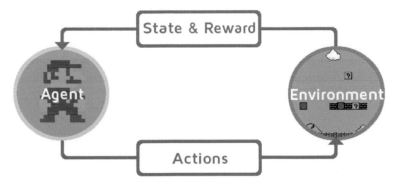

Figure 9.1 Reinforcement learning.
Credit: https://www.freecodecamp.org/news/a-brief-introduction-to-reinforcement-learning-7799af5840db/.

Spatial navigation, discussed in Chapter 7, is originally developed for essential functions such as finding food, water and mates. This has been shown in monkeys and humans to include navigation in abstract space[31, 32]. Various patterns are learned and updated every day. Replay/preplay by the claustrum-MEC (Medial entorhinal cortex) cells may be more for vector navigation or planning, and replay by the cortex-hippocampal place cells may be more for memory and accurate localization. Alternative routes are replayed at decision points[33] (Fig. 9.2).

The brain prepares itself for the future, while asleep and while awake (Chapter 6). When disengaged from the ongoing task, replay/preplay is less focused, activating a myriad of "places" and "paths." Animals with eyes at the front, such as monkeys and humans, can use the two-dimensional (2D) grids of different sizes for gazing at images[34, 35] and perhaps reading (Section 8.7). Social relations[36, 37] (Section 7.9), and likely other schemes, are also projected onto such a 2D-map. Imagination could simply mean to preplay within proper frameworks (not as mere successors of a previous step[38, 39]). And as we noted in Chapter 6, dreaming associates with visuo-spatial skills in children[40].

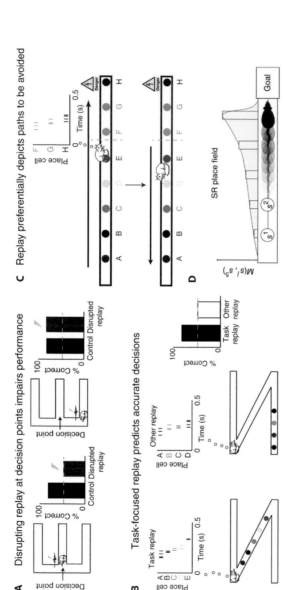

Figure 9.2 Planning the path and pinpointing the goal. (A) Disrupting sharp-wave ripples at decision points in a spatial alternation task ("W maze") was associated with impaired performance compared to control animals (left). When sharp-wave ripples were disrupted at nondecision points performance was preserved (right)[41]. (B) Replay was recorded at the corners of a Z-shaped track preceding correct and incorrect turns. When replay depicted positions consistent with the animals' current positions (for example, proximal locations, left) the rats were more likely to make the correct turn (right). Whereas if replay depicted positions not immediately relevant to current behavior (middle), animals were less likely to make the correct turn (right)[33]. (C) Following training on an inhibitory avoidance task (learning to associate the end of a linear runway with a foot shock), replay during pauses preceding entry to a shock zone preferentially depicted paths towards the feared zone (top) and was associated with the animals turning away from the shock zone and running in the opposite direction (bottom)[42]. (D) The place field for a single successor representation (SR)-encoding cell skews backward toward past states that predict the cell's preferred state. The receptive field $M(\cdot, s^5)$ illustrated here corresponds to a column of the SR matrix.

Credit: Part A, B, C from Fig. 5 of ref. 43; Part D from Fig. 2e of ref. 38.

As curious human beings, our cognitive exploration is essentially also a navigation towards a goal (Figs. 7.12 and 7.13). Or, more explorative in a large open field, as in a Lévy walk, the direction may occasionally shift, in a balance between a reward of expected size and a potential larger reward (Figs. 9.2 and 9.3). Stochastic steps can be generated by grid cells[44]. To make a turn does not necessarily mean abandoning a direction (Figs. 9.3 and 9.4)—the hippocampus is still keeping track despite the turn of the head, so the previous point could still be active enough for a future comeback and further exploration.

Some navigational training for the MEC grid cells might be a way of not being emersed in acetylcholine-driven cortical-hippocampal memory replays at an old age (e.g., Figs. 6.10, and 5.14). For individuals who have not been rewarded for some types of activities in the past, more positive experiences could still promote some new cells and new circuits that get incorporated into existing networks. The tendency to spend too much in the digital world also lies in a lack of concrete references in setting loss versus reward.

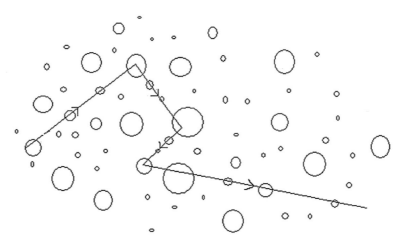

Figure 9.3 Schematic representation of the generation of Lévy walk movement patterns with Lévy exponent $\mu = 2$ in a landscape containing patchily distributed resources. Direction of travel is indicated by arrows and the initial direction of travel was chosen at random. Foragers continue to move in the same direction until they encounter a patch (circle) whose quality (illustrated by the patch size) is perceived to be greater than that of the first patch encountered along that direction of travel. A new direction of travel is then chosen at random and the cycle repeats.

Credit: Fig. 5 of ref. 45.

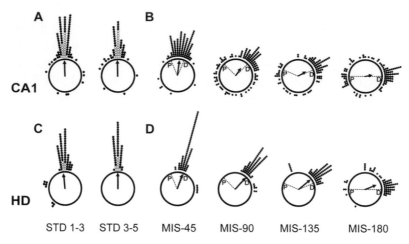

Figure 9.4 Splitting of hippocampal CA1 place field in the presence of conflicting cues, while head direction cells maintain an internal coherence. Each dot on the polar plot (open dots, 10 data points; filled dots, 1 data point) represents the amount of rotation of a CA1 place field and a head direction cell tuning curve between two standard sessions (A, C) and between standard (STD) versus different mismatch (MIS) sessions (45°, 90°, 135°, and 180°) (B, D). The arrows represent the mean angle of rotation of the population, and the length of the arrow represents the angular dispersion around the mean. The dashed lines indicate the rotation amounts of the proximal (CCW) and distal (CW) cue sets. Because cells were recorded over multiple sessions and days, each dot does not correspond to a unique cell. P, Proximal cue; D, distal cue.

Credit: Fig. 5 of ref. 46.

9.2 Expected or Unexpected

For rats and mice who were born to regard confined spaces as safe, an electric shock in the feet in some branch of a maze is both unpleasant and unexpected enough to be remembered, with new dendrites or new neurons to link the place and the incident. One or two strong synapses in each neuron would be sufficient[47]. To be able to predict and avoid future incidences, clues will be stored and reinforced by experience (Figs. 9.1 and 9.5), and notable differences between events are also stored[48] (Section 5.5). I do not think replays are there to prevent catastrophic forgetting[49]; Rather, a major function of the hashed replays/preplays would be to simulate all the likely important possibilities (Section 9.3), properly allocate the excitatory and inhibitory neurons to the important tasks[15, 50], and store new information at the proper places (Section 7.8). The balance between forward replay and reverse replay sets the phase of θ waves[29, 30, 38], which involve interneurons together with pyramidal neurons[48] (Section 6.3).

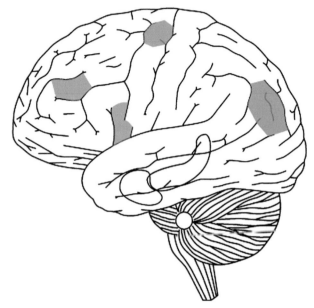

Figure 9.5 Hypothetical use of the iterative GAN (Generative adversarial network) discussed in Section 7.8 to project grids to different positions of the cortex. The conscious shift in attention perhaps require projection from the claustrum to the MEC, which then with detailed indices in the hippocampus (Fig. 4.11), finds appropriate patterns to be used at the cortex.

Credit: Changxing Su, Huijue Jia.

We mentioned analog representation of numbers (Chapter 8) and features of light stripes (Fig. 4.2). This information is stored with an orientation (Figs. 9.4, 9.5, and 9.6), and the extent of neuronal firing possibly scales with the θ (and γ) phase projection in that particular orientation (e.g., ref. 47), which could be tweaked with activity. The grid structure helps predict the subsequent steps[38, 39]. Rules are stored as vectors with their own local maps, and each event is hashed against an existing set of patterns (Fig. 9.7). If things align nicely, it would be satisfactory, and more likely utilized again in the future (large dendritic spines do not easily get larger, Chapter 5). Alignment at the 30° low activity points might lead to no adjustment (Fig. 9.6), one of the potential explanations for the lack of improvement in model fitting despite further training (e.g., Fig. 2.8B) in addition to insufficient time for neuronal growth.

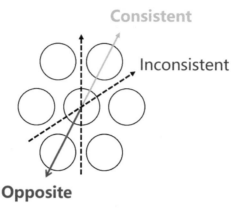

Figure 9.6 A hexagonal grid used to judge how consistent a new event is relative to an existing vector in the brain. The circles can represent one or more neuronal cells in the cortex. Like the grid cell activities during spatial navigation (Chapter 6), the solid arrow indicates a perfect alignment, while the dashed arrows mark the lowest activities at 30° angle. An event that is opposite to the existing vector should induce a strong activity.

Credit: Huijue Jia.

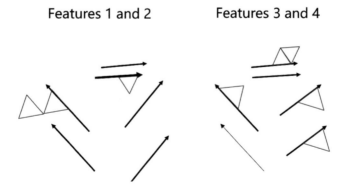

Figure 9.7 Sets of vectors that a new event is compared against. Consistent alignment to certain vectors should increase their encoded probability. Inconsistencies would be properly stored in new cells or new branches of existing cells (Chapter 4).

Credit: Huijue Jia.

Greater activity would be induced for unexpected results, to recruit dendrites and neurons that can encode the novelty. To refer to familiar examples and to setup something unexpected are commonly used tricks in education and in presentation. The familiarity likely provides reference points for comparison and information storage (Figs. 9.5 and 9.7), and the unexpectedness creates a strong signal (Fig. 9.6, Chapter 4, Fig. 4.7) to increase the likelihood that new spines, new branches and new cells are recruited to the site. Results are thus prioritized, so that a complete picture quickly emerges (though with uncertainty, Section 5.2), without a lot of training data.

The entorhinal-hippocampal grids might be projected onto the neocortex where appropriate (Figs. 9.5, 9.8, and 6.11), which searches for existing information, both sensory and conceptual (e.g., refs. 31, 51), and generates a mental representation (Figs. 4.11, 7.15, and 9.5). With one or more points anchored, this would be a more flexible version of transfer learning, which completes the pattern according to stored patterns that come out of the searching process and are combined in a new sequence.

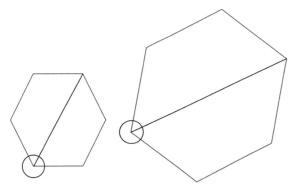

Figure 9.8 Grids of different sizes and orientations, anchored at one existing point. More on memory, forgetting and spatial-temporal orientations in Chapters 4, 5, 6.

Credit: Huijue Jia.

With previous experience on the "distance" between events (e.g., an existing pattern of hippocampal connections to be utilized at the new places, Chapter 4) and with replay/preplay (Chapter 5), branches of existing neurons may be able to hold a place of a certain size for details to be filled in (Figs. 9.8 and 9.9). Starting with a large grid and refining with one or more smaller grids[44, 52] (Fig. 7.3). Hippocampal index words and sensory input from multiple categories can also help define the location. Otherwise the inaccuracy would increase with more steps from a known point (Fig. 9.2D; Chapter 7).

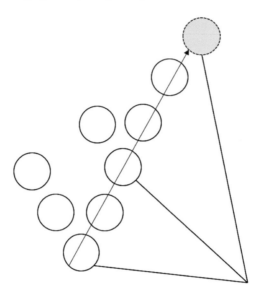

Figure 9.9 A series of related steps that are being worked out in the brain. The current point of interest is highlighted in yellow, which may shift as in Fig. 9.2, or become more solid with concrete information.

Credit: Huijue Jia.

It is uniquely human that knowledge is systematically passed on from people whom we have never met, so that we can accumulate useful patterns (e.g., someone's life hundreds of years ago) beyond the spatiotemporal scale that an individual animal could ever experience.

9.3 Path Diagrams and Counterfactuals in View of Navigation

For causal reasoning[53, 54], path diagrams (e.g., Fig. 9.10, or signaling pathways in molecular biology) are nicely named to reflect its similarity to spatial navigation (Fig. 9.11). Each event is a node, just like how it is possibly stored by hippocampal pyramidal cells through interaction with the cortex (Section 4.4). Some get activated more easily than others (including hallucinating about someone calling one's name[55]). A shorter and more general path would be more efficient and more generalizable, saving cells (e.g., grid cells vs. place cells, Fig. 7.4; right hippocampal "place cells" for coherent event boundaries in storytelling[56]) and accumulate higher weights with each successful run (Chapter 5).

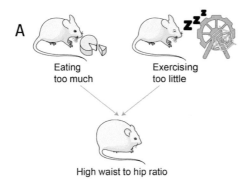

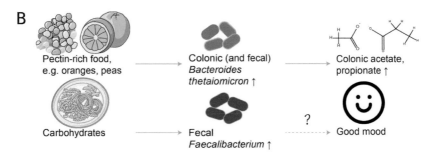

Figure 9.10 Illustrated examples of path diagrams. (A) A collider at getting obese. (B) Two examples of likely causal paths from food, gut microbe, to human physiology.

Credit: Fig. 6.3 of Chapter 6 of ref. 58 on causality.

In the existing framework of probable routes (e.g., Fig. 9.11), to make a counterfactual is equivalent to asking whether one could still reach a goal location, when a path is disabled on the mental representation of the map. But the answer also relates to the alternative paths (as in the allocation of Bayesian probabilities and scientific investigations[53, 54]) and the current weight of the disabled path on the map (prior information that guides the choice of patterns and their existing probabilities), after confirming that the path existed.

Figure 9.11 All roads lead to Rome. Counterfactuals essentially mean to remove a path and estimate whether we can still reach a target place. If there is only one major path according to our prior knowledge, there would be no question about causality. Each individual has his or her own mental map, but commonsense knowledge would be stored in short paths with heavy weights in most people.

Credit: https://www.istockphoto.com/photo/all-roads-lead-to-rome-the-map-gm182160878-19950566.

Coincidence detection in individual neurons with input from elsewhere (e.g., refs. 5, 57) could be a way to gate a path with a second probability. More than one synapse on the same dendritic branch (which could come from parallel circuits with a delay in between) or even on a single dendritic spine might be involved (Figs. 5.6 and 6.7).

The maps are inevitably different in each individual brain, while consensus could still be reached. An overlooked path may turn out to be gateway to something important.

9.4   Summary

The feedback hashing model for searching and memorizing information (Chapters 4 and 5), tidied up by sleep every night (Chapter 6), leads to the more iterative and adaptive GAN than existing algorithms (Fig. 7.15). After a quick search, loosely matching patterns learned previously, might be projected to the cortex through the claustrum-EC-hippocampal waves (Figs. 9.5, 9.7, and 9.8), to guide behavior in a new situation. So it is never an entirely new situation for an adult, but a loose match or a new combination would likely encourage more improvising.

Following a right path to a reward, or avoiding a wrong path to something bad, is an everyday practice of reinforcement learning for an animal. As advocated by Dr. Judea Pearl, the brain has an intrinsic tendency for causal relationships[53, 54]. The tentative associations may be lacking detailed paths (mechanism), but the Bayesian brain quickly accumulates weights in dendritic spines for each rule that has worked (Chapter 5). When interventions such as randomized controlled trials are trying to establish a path, counterfactuals mentally block a path and ask whether we can still get there. The map is in everyone's brain, so if we can no longer get to the spot, we know it.

I try not to say too much at the end of this book. Please enjoy navigating!

Questions

1. When a previously successful pattern is mobilized for the current task, how frequently shall we check for discrepancies in the GAN and tweak the parameters? When a second or third pattern is in waiting, how does the allocation of resources change overtime?

2. How should the reward sizes and frequency be set, to encourage venturing into something that has previously been unsuccessful?

3. Struck by something unexpected, how do you think the interneurons and pyramidal neurons work differently during reverse replay as opposed to forward replay[29] (Section 5.5)? And what about their synapses?

4. There is some cost to the network when a more general association is replaced by a causal path, what would be a healthy ratio of association and causality, for subnetworks of different sizes and functions?

References

1. Kraft, T. S. *et al.* The energetics of uniquely human subsistence strategies. *Science (80-.).* **374** (2021).

2. Tooley, U. A., Bassett, D. S. & Mackey, A. P. Environmental influences on the pace of brain development. *Nat. Rev. Neurosci.* **22**, 372–384 (2021).

3. Braitenberg, V. & Schüz, A. Cortex: statistics and geometry of neuronal connectivity. *Cortex Stat. Geom. Neuronal Connect* (1998) doi:10.1007/978-3-662-03733-1.

4. Shepherd, G. M. G. & Yamawaki, N. Untangling the cortico-thalamo-cortical loop: cellular pieces of a knotty circuit puzzle. *Nat. Rev. Neurosci.* **22**, 389–406 (2021).

5. Larkum, M. A cellular mechanism for cortical associations: an organizing principle for the cerebral cortex. *Trends Neurosci.* **36**, 141–51 (2013).

6. Sharpe, M. J. *et al.* Dopamine transients are sufficient and necessary for acquisition of model-based associations. *Nat. Neurosci.* **20**, 735–742 (2017).

7. Hage, T. A. *et al.* Synaptic connectivity to L2/3 of primary visual cortex measured by two-photon optogenetic stimulation. *Elife* **11** (2022).

8. Motta, A. *et al.* Dense connectomic reconstruction in layer 4 of the somatosensory cortex. *Science (80-.).* **366**, eaay3134 (2019).

9. Sirota, A. *et al.* Entrainment of neocortical neurons and gamma oscillations by the hippocampal theta rhythm. *Neuron* **60**, 683–97 (2008).

10. G, D. *et al.* Perirhinal input to neocortical layer 1 controls learning. *Science* **370**, eaaz3136 (2020).

11. Bush, D. *et al.* Human hippocampal theta power indicates movement onset and distance travelled. *Proc. Natl. Acad. Sci.* **114**, 12297–12302 (2017).

12. Oberto, V. J. *et al.* Distributed cell assemblies spanning prefrontal cortex and striatum. *Curr. Biol.* **32**, 1–13.e6 (2022).

13. Blackman, A. V., Abrahamsson, T., Costa, R. P., Lalanne, T. & Sjöström, P. J. Target-cell-specific short-term plasticity in local circuits. *Front. Synaptic Neurosci.* **5**, 11 (2013).

14. Lagler, M. *et al.* Divisions of identified parvalbumin-expressing basket cells during working memory-guided decision making. *Neuron* **91**, 1390–1401 (2016).

15. Terada, S. *et al.* Adaptive stimulus selection for consolidation in the hippocampus. *Nature* **601**, 240–244 (2022).

16. Sadeh, S. & Clopath, C. Excitatory-inhibitory balance modulates the formation and dynamics of neuronal assemblies in cortical networks. *Sci. Adv.* **7**, eabg8411 (2021).

17. Adamantidis, A. R., Gutierrez Herrera, C. & Gent, T. C. Oscillating circuitries in the sleeping brain. *Nat. Rev. Neurosci.* **20**, 746–762 (2019).

18. Geiller, T. *et al.* Large-scale 3D two-photon imaging of molecularly identified CA1 interneuron dynamics in behaving mice. *Neuron* **108**, 968–983.e9 (2020).

19. Valero, M., Zutshi, I., Yoon, E. & Buzsáki, G. Probing subthreshold dynamics of hippocampal neurons by pulsed optogenetics. *Science (80-.)*. **375**, 570–574 (2022).

20. Wu, Y. E. *et al.* Neural control of affiliative touch in prosocial interaction. *Nature* **599**, 262–267 (2021).

21. Williams, T. B. *et al.* Testing models at the neural level reveals how the brain computes subjective value. *Proc. Natl. Acad. Sci.* **118**, e2106237118 (2021).

22. Sosa, M. & Giocomo, L. M. Navigating for reward. *Nat. Rev. Neurosci.* **22**, 472–487 (2021).

23. Liu, C., Goel, P. & Kaeser, P. S. Spatial and temporal scales of dopamine transmission. *Nat. Rev. Neurosci.* **22**, 345–358 (2021).

24. Wang, J. X. *et al.* Prefrontal cortex as a meta-reinforcement learning system. *Nat. Neurosci.* **21**, 860–868 (2018).

25. Cheng, S. & Frank, L. M. New experiences enhance coordinated neural activity in the hippocampus. *Neuron* **57**, 303–13 (2008).

26. Foster, D. J. & Knierim, J. J. Sequence learning and the role of the hippocampus in rodent navigation. *Curr. Opin. Neurobiol.* **22**, 294–300 (2012).

27. Ambrose, R. E., Pfeiffer, B. E. & Foster, D. J. Reverse replay of hippocampal place cells is uniquely modulated by changing reward. *Neuron* **91**, 1124–1136 (2016).

28. Singer, A. C. & Frank, L. M. Rewarded outcomes enhance reactivation of experience in the hippocampus. *Neuron* **64**, 910–21 (2009).

29. Wang, M., Foster, D. J. & Pfeiffer, B. E. Alternating sequences of future and past behavior encoded within hippocampal theta oscillations. *Science* **370**, 247–250 (2020).

30. Schapiro, A. C., Turk-Browne, N. B., Botvinick, M. M. & Norman, K. A. Complementary learning systems within the hippocampus: a neural network modelling approach to reconciling episodic memory with statistical learning. *Philos. Trans. R. Soc. B Biol. Sci.* **372**, 20160049 (2017).

31. Constantinescu, A. O., O'Reilly, J. X. & Behrens, T. E. J. Organizing conceptual knowledge in humans with a gridlike code. *Science* **352**, 1464–1468 (2016).

32. Knudsen, E. B. & Wallis, J. D. Hippocampal neurons construct a map of an abstract value space. *Cell* **184**, 4640–4650.e10 (2021).

33. Ólafsdóttir, H. F., Carpenter, F. & Barry, C. Task demands predict a dynamic switch in the content of awake hippocampal replay. *Neuron* **96**, 925–935.e6 (2017).

34. Killian, N. J., Jutras, M. J. & Buffalo, E. A. A map of visual space in the primate entorhinal cortex. *Nature* **491**, 761–764 (2012).

35. MLR, M. & EA, B. Neurons in primate entorhinal cortex represent gaze position in multiple spatial reference frames. *J. Neurosci.* **38**, 2430–2441 (2018).

36. Tavares, R. M. *et al.* A Map for social navigation in the human brain. *Neuron* **87**, 231–43 (2015).

37. Park, S. A., Miller, D. S. & Boorman, E. D. Inferences on a multi-dimensional social hierarchy use a grid-like code. *Nat. Neurosci.* **24**, 1292–1301 (2021).

38. Stachenfeld, K. L., Botvinick, M. M. & Gershman, S. J. The hippocampus as a predictive map. *Nat. Neurosci.* **20**, 1643–1653 (2017).

39. Whittington, J. C. R. *et al.* The Tolman-Eichenbaum Machine: unifying space and relational memory through generalization in the hippocampal formation. *Cell* **183**, 1249–1263.e23 (2020).

40. Nir, Y. & Tononi, G. Dreaming and the brain: from phenomenology to neurophysiology. *Trends Cogn. Sci.* **14**, 88–100 (2010).

41. Jadhav, S. P., Kemere, C., German, P. W. & Frank, L. M. Awake hippocampal sharp-wave ripples support spatial memory. *Science* **336**, 1454–8 (2012).

42. Wu, C.-T., Haggerty, D., Kemere, C. & Ji, D. Hippocampal awake replay in fear memory retrieval. *Nat. Neurosci.* **20**, 571–580 (2017).

43. Ólafsdóttir, H. F., Bush, D. & Barry, C. The role of hippocampal replay in memory and planning. *Curr. Biol.* **28**, R37–R50 (2018).

44. Stemmler, M., Mathis, A. & Herz, A. V. M. Connecting multiple spatial scales to decode the population activity of grid cells. *Sci. Adv.* **1**, e1500816 (2015).

45. Reynolds, A. Liberating Lévy walk research from the shackles of optimal foraging. *Phys. Life Rev.* **14**, 59–83 (2015).

46. Yoganarasimha, D., Yu, X. & Knierim, J. J. Head direction cell representations maintain internal coherence during conflicting proximal and distal cue rotations: comparison with hippocampal place cells. *J. Neurosci.* **26**, 622–31 (2006).

47. L, G., A, R. & M, H. Active dendrites enable strong but sparse inputs to determine orientation selectivity. *Proc. Natl. Acad. Sci. U. S. A.* **118**, e2017339118 (2021).

48. Hertäg, L. & Clopath, C. Prediction-error neurons in circuits with multiple neuron types: Formation, refinement, and functional implications. *Proc. Natl. Acad. Sci.* **119**, e2115699119 (2022).

49. Roscow, E. L., Chua, R., Costa, R. P., Jones, M. W. & Lepora, N. Learning offline: memory replay in biological and artificial reinforcement learning. *Trends Neurosci.* **44**, 808–821 (2021).

50. Rolotti, S. V. *et al.* Local feedback inhibition tightly controls rapid formation of hippocampal place fields. *Neuron* **110**, 783–794.e6 (2022).

51. Bao, X. *et al.* Grid-like neural representations support olfactory navigation of a two-dimensional odor space. *Neuron* **102**, 1066–1075. e5 (2019).

52. Bush, D., Barry, C., Manson, D. & Burgess, N. Using grid cells for navigation. *Neuron* **87**, 507–20 (2015).

53. Pearl, J. & Mackenzie, D. *The Book of Why* (Basic Books, 2018).

54. Jaynes, E. T. *Probability Theory: The Logic of Science* (Cambridge University Press, 2003).

55. Horga, G. & Abi-Dargham, A. An integrative framework for perceptual disturbances in psychosis. *Nat. Rev. Neurosci.* **20**, 763–778 (2019).

56. Cohn-Sheehy, B. I. *et al.* The hippocampus constructs narrative memories across distant events. *Curr. Biol.* **31**, 4935–4945.e7 (2021).

57. Egger, R. *et al.* Cortical output is gated by horizontally projecting neurons in the deep layers. *Neuron* **105**, 122–137.e8 (2020).

58. Jia, H. *Investigating Human Diseases with the Microbiome: Meta-genomics Bench to Bedside.* (Academic Press, 2022).

Index